Manufacturing Systems Engineering

Manufacturing Systems Engineering

Contributors

Yihai He, Zhenzhen He et al.

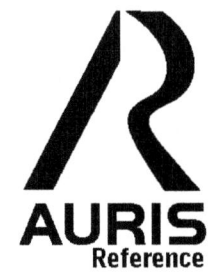

AURIS
Reference

www.aurisreference.com

Manufacturing Systems Engineering

Contributors: Yihai He, Zhenzhen He et al.

Published by Auris Reference Limited
www.aurisreference.com

United Kingdom

Manufacturing Systems Engineering

ISBN: 978-1-78154-933-9

British Library Cataloguing in Publication Data
A CIP record for this book is available from the British Library

Printed in the United Kingdom

Exclusively distributed by CBS Publishers & Distributors Pvt. Ltd.

Sales & Distribution Rights only for India, Pakistan, Bangladesh, Sri Lanka, Nepal and Bhutan. This book is not to be sold outside these territories.

Contents

List of Abbreviations

AV	Antithetic Variables
ANCs	average number of customers
AWT	average waiting time
CBR	Case-Based Reasoning systems
CRVs	Common Random Variables
CF	Compact Framework
CTL	Computation Tree Logic
CNC	computer numerically controlled
DEA	Data Envelopment Analysis
DLD	Direct Laser Deposition
EF	Economical fabrication
ERP	enterprise resource planning
ES	Expert Systems
Ectl	extended Computation Tree Logic
FMEA	failure modes and effects analyses
GMAW	Gas metal arc welding
GUI	Graphical User Interface
HDPE	High density polyethylene
HLA	High Level Architecture
IT	Information technology
IPD	Integrated Product Development
IMS	Intelligent Manufacturing Systems
KQCs	Key product characteristics
LAMP	laser aided manufacturing process
LC	Laser Consolidation
LBAM	Laser-Based Additive Manufacturing
LEDs	light emitting diodes
LLDPE	linear low density polyethylene
MP	Manufacturing process
MP	Programming models
NN	Neural Networks
PCs	Personal computers
PM	preventive maintenance
PCB	printed circuit board
PDS	product design specification
PAC	Programmable Automation Controllers
REMS	reconfigurable and energy-efficient manufacturing system
R-TNCESs	reconfigurable timed net condition event systems
SS	Stratified Sampling
SVM	support vector machines
VR	variance reduction
VRTs	variance reduction techniques
WIP	Work in Process

List of Contributors

Yihai He
School of Reliability and Systems Engineering, Beihang University, Beijing 100191, China
Department of Systems Engineering and Engineering Management, City University of Hong Kong, Kowloon, Hong Kong

Zhenzhen He
School of Reliability and Systems Engineering, Beihang University, Beijing 100191, China

Linbo Wang
School of Reliability and Systems Engineering, Beihang University, Beijing 100191, China

Changchao Gu
School of Reliability and Systems Engineering, Beihang University, Beijing 100191, China

Jiafeng Zhang,
School of Electro-Mechanical Engineering, Xidian University, Xi'an 710071, China
Chair of Automation and Energy Systems, Saarland University, 66123 Saarbrücken, Germany

Mohamed Khalgui
University of Carthage, 1054 Carthage, Tunisia

Wassim Mohamed Boussahel
Chair of Automation and Energy Systems, Saarland University, 66123 Saarbrücken, Germany
Zentrum für Mechatronik und Automatisierungstechnik, 66121 Saarbrücken, Germany

Georg Frey
Chair of Automation and Energy Systems, Saarland University, 66123 Saarbrücken, Germany

ChiTin Hon
Institute of Systems Engineering, Macau University of Science and Technology, Taipa, Macau

Naiqi Wu
Institute of Systems Engineering, Macau University of Science and Technology, Taipa, Macau

Zhiwu Li
Institute of Systems Engineering, Macau University of Science and Technology, Taipa, Macau
Faculty of Engineering, King Abdulaziz University, Jeddah 21589, Saudi Arabia

R. A. R. C. Gopurane
Department Mechanical Engineering, University of Moratuwa, Kaubedda, Sri Lanka

T. S. S. Jayawardane
Department of Textile and Clothing Technology, University of Moratuwa, Kaubedda, Sri Lanka

Ergün Eraslan
Department of Industrial Engineering, Baskent University, Eskisehir Road 22.km, 06590 Ankara, Turkey

Berna Dengiz
Department of Industrial Engineering, Baskent University, Eskisehir Road 22.km, 06590 Ankara, Turkey

Xun Gong,
State Key Lab of Fluid Power Transmission and Control, Zhejiang University, Hangzhou 310027, China
Robotics and Microsystems Center, Soochow University, Suzhou 215006, China

Yixiong Feng
State Key Lab of Fluid Power Transmission and Control, Zhejiang University, Hangzhou 310027, China

Hao Zheng
State Key Lab of Fluid Power Transmission and Control, Zhejiang University, Hangzhou 310027, China

Jianrong Tan
State Key Lab of Fluid Power Transmission and Control, Zhejiang University, Hangzhou 310027, China

Aslı Aksoy
Uludag University Department of Industrial Engineering, Bursa Turkey

Nursel Öztürk
Uludag University Department of Industrial Engineering, Bursa Turkey

Tomasz Mączka
Rzeszów University of Technology Poland

Tomasz Żabiński
Rzeszów University of Technology Poland

Jacquelyn K. S. Nagel
James Madison University USA

Frank W. Liou
Missouri University of Science and Technology USA

Hasse Nylund
Tampere University of Technology Finland

Paul H Andersson
Tampere University of Technology Finland

Michael A. Saliba
Department of Industrial and Manufacturing Engineering, University of Malta Malta

Anthony Caruana
Department of Industrial and Manufacturing Engineering, University of Malta Malta

Preface

Providing a description of some of the phenomena in material flow in manufacturing systems, the text *Manufacturing Systems Engineering* details some of the potentially disruptive events that affect the production process. Manufacturing technology is concerned with the flow of materials from the acquisition of raw materials, through conversion in the workshop to the shipping of finished goods to the customer. First chapter presents an approach to model the manufacturing system reliability dynamically based on their operation data of process quality and output data of product reliability. In second chapter, we deal with the formal modeling and verification of reconfigurable and energy-efficient manufacturing systems (REMSs) that are considered as reconfigurable discrete event control systems. Third chapter presents preliminary analysis, modeling and simulation strategies of a poly-bag manufacturing system. In fourth chapter, the effect of control variates (CVs) and Stratified Sampling (SS) techniques in reducing variance of the performance measurements of M/M/1 and GI/G/1 queue models has been investigated considering four probability distributions utilizing randomly generated parameters for arrival and service processes. An adaptive maintenance model oriented to process environment of the manufacturing systems has been presented in fifth chapter. Sixth chapter provides a basic definition of global outsourcing and analyzes global outsourcing as either an opportunity or a threat. Seventh chapter focuses on platform for intelligent manufacturing systems with elements of knowledge discovery. The goal of eighth chapter is to summarize the key research findings related to the design, development, and integration of a hybrid manufacturing process that utilizes laser metal deposition (LMD) to produce fully dense, finished metallic parts. Ninth chapter discusses on the challenges and opportunities of digital manufacturing supporting the decision making in autonomous and collaborative actions of manufacturing companies. Last chapter addresses the situation where a manufacturer needs to investigate a potential manufacturing system migration for the assembly of a part family of products, where no or minimal product design changes are allowed.

Chapter 1

RELIABILITY MODELING AND OPTIMIZATION STRATEGY FOR MANUFACTURING SYSTEM BASED ON RQR CHAIN

Yihai He,[1,2] Zhenzhen He,[1] Linbo Wang,[1] and Changchao Gu[1]

[1]School of Reliability and Systems Engineering, Beihang University, Beijing 100191, China
[2]Department of Systems Engineering and Engineering Management, City University of Hong Kong, Kowloon, Hong Kong

ABSTRACT

Accurate and dynamic reliability modeling for the running manufacturing system is the prerequisite to implement preventive maintenance. However, existing studies could not output the reliability value in real time because their abandonment of the quality inspection data originated in the operation process of manufacturing system. Therefore, this paper presents an approach to model the manufacturing system reliability dynamically based on their operation data of process quality and output data of product reliability. Firstly, on the basis of importance explanation of the quality variations in manufacturing process as the linkage for the manufacturing system reliability and product inherent reliability, the RQR chain which could represent the relationships between them is put forward, and the product qualified probability is proposed to quantify the impacts of quality variation in manufacturing process on the reliability of manufacturing system further. Secondly, the impact of qualified probability on the product inherent reliability is expounded, and the modeling approach of manufacturing system reliability based on the qualified probability is presented. Thirdly, the preventive maintenance optimization strategy for manufacturing system driven by the loss of manufacturing quality variation is proposed. Finally, the validity of the proposed approach is verified by the reliability analysis and optimization example of engine cover manufacturing system.

INTRODUCTION

To meet the demands of high reliability and long life of the product, integrated analysis, assurance, and optimization for reliability are required to be carried out in the lifecycle of design, manufacture, and usage. However, for a long time, most of traditional reliability studies had merely focused on the design and usage stages, and reliability technology suitable for the manufacturing process has always been ignored, lacking proper attention it deserved, which caused the serious degradation of product reliability after batch production frequently, and resulting in a high infant failure rate, and the product inherent reliability could not meet the increasingly stringent design reliability requirements [1]. As we all know, product is the output of the manufacturing process which is the implementation form of the manufacturing system. Therefore, the reliability of final produced product is closely related to the reliability of manufacturing system and the quality of manufacturing process. Usually, even a good design cannot guarantee that the manufactured products achieve the satisfactory reliability when the design quality of manufacturing system is poor [2]. Thus, it can be seen that integrating the reliability modeling and optimization for manufacturing system is crucial to ensure the product reliability.

Practices have proved that uncertain factors like quality variation in manufacturing process and deteriorations of system components could lead to the degradation of manufacturing system, which should affect the quality of manufacturing process interactively. And when quality variations are cumulated and amplified, the number of potential defects of products is arising, which would trigger the decline of product inherent reliability finally. In order to minimize the decline of product inherent reliability with respect to the design specifications, identifying and optimizing the critical factors in manufacturing that contribute to the product reliability degradation systematically are becoming the research focus of reliability engineering currently, and how to carry out product reliability oriented reliability modeling and optimization of manufacturing system is the most urgent and task.

At different nodes of product life cycle, product reliability exhibits different characteristics. Murthy [3] defined the evolution chain which transfers product reliability from design, manufacturing, transportation, sale, and usage, enriching notation of product reliability at different stages, and named the reliability in manufacturing as product inherent reliability. Then, Jiang and Murthy [4] pointed out the negative impact of variations on the reliability during the product life cycle via the transmission chain and pointed out that the quality variations and assembly errors are the root reasons causing the deterioration of product inherent reliability. As to product inherent reliability in manufacturing, Li et al. [5, 6] noted that both the reliability of manufacturing system and quality of manufacturing process are the critical roles to ensure and improve the product quality and reliability. Inman et al. [7] believed that performance of manufacturing system severely restricted product quality and reliability, and upgrading the manufacturing equipment or adjusting the technological process could promote the ability of manufacturing system as well as ensuring and optimizing product quality and reliability. To some extent, the ability of trouble free operation of manufacturing system determines the level of inherent reliability formed in manufacturing process.

Traditional reliability modeling of manufacturing system tends to follow the classic reliability block diagram method, fault tree analysis, Petri nets, and so forth, which caused a comprehensive analysis and dynamical assessment for manufacturing system to be complex or inconvenient. Based on the data of system operation and maintenance, Li and Ni [8] used the maximum likelihood estimation method to estimate the reliability of manufacturing system, which provided the basis for carrying out preventive maintenance of manufacturing system. Lin and Chang [9] proposed the limited manufacturing network model, and after mining operating failures and rework data, an analysis model of manufacturing system reliability was established. Li et al. [10] created a prediction model of manufacturing system using the grey model, and the author stated that the weaknesses of manufacturing system could be identified by the proposed model. Considering the plenty of quality data existing in manufacturing process, Chen and Jin [11, 12] put forward a Quality-Reliability chain model based on the interaction between manufacturing process quality and manufacturing system reliability, and the reliability analysis and maintenance optimization of manufacturing system were expounded based on the proposed Quality-Reliability chain. Zhang et al. [13] presented a reliability modeling approach of manufacturing system using dimensions of process quality. Rafiee et al. [14] analyzed four typical vibration modes and their effects on the degradation rate of manufacturing process and modeled the complex manufacturing system reliability like MEMS and so on. Regarding the maintenance strategy of manufacturing systems, Li et al. [15] investigated the economic production quantity model jointly considering product deterioration and proposed an EPQ (economical production quantity) model for deteriorating production system and items with rework. Gong et al. [16] explored an adaptive maintenance model of the process environment to diagnose the progressive faults in manufacturing systems. Tlili et al. [17] proposed a new modeling approach based on the fact that the degradation process is modeled by the wiener process. Hajej et al. [18–21] studied integrated maintenance strategies and policies jointly considering the optimization problems of subcontracting, product returns, lease contract, and so forth, which provide a solid foundation to develop the integrated maintenance strategies optimization for manufacturing system. Mifdal et al. [22] presented a joint optimization approach of maintenance and production planning for a multiple-product manufacturing system, which could establish sequentially an economical production plan and an optimal maintenance strategy considering the influence of the production rate on the system's degradation.

As can be seen from the above literature analysis, the interaction between product inherent reliability and manufacturing system reliability is not defined, and studies on the product reliability oriented reliability modeling and optimization of manufacturing system are few, which are in urgent need to develop reliability assurance in manufacturing. Therefore, in order to promote the joint reliability optimization of manufacturing system and the produced product, an integrated model named RQR chain is proposed by extending the Quality-Reliability chain in this paper, the RQR chain could describe the coeffects of the manufacturing system reliability R_m, manufacturing process quality Q_p, and product inherent reliability Rp, and the impact of manufacturing system reliability on the product inherent reliability is expounded specifically based on product qualified probability. At the same time, the quantitative analysis model and optimization strategies of manufacturing system reliability are given. Comparing to previous related studies in the frame of integrated reliability and maintenance optimization of manufacturing system, the main contributions of this paper are as follows:

(a)   RQR chain is brought forth and the product qualified probability is proposed to quantify the impacts of manufacturing process quality on the manufacturing system reliability for the first time. The product qualified probability driven RQR chain could make integrated reliability optimization of manufacturing system and produced product possible.

(b)   A reliability modeling approach of manufacturing system based on the proposed qualified probability is presented. The impact of qualified probability on the inherent reliability of produced product oriented is expounded at the first time in this paper, and the concept of the reliability of manufacturing is extended by including the requirement of product quality that is qualified by time in this proposed approach.

(c)   The preventive maintenance optimization strategy for manufacturing system driven by the loss of manufacturing quality variation is proposed by the aid of the product qualified probability. The proposed product qualified probability based optimization approach could make the real time preventive maintenance possible when the abnormal quality variations occurred in the produced work pieces in the batch production.

The rest of the paper is organized as follows. Section 2 emphasizes the role of manufacturing process quality variations as a transfer bridge for analyzing the influence of manufacturing system reliability on product inherent reliability, and the RQR chain based on product qualified probability is put forward. Modeling of manufacturing system reliability based on the qualified probability is analyzed in Section 3. With reference to the results of the proposed model, Section 4 presents some optimization strategies driven by product quality loss. Section 5 discusses the application mode and the effects of the proposed method in an automotive cylinder head manufacturing system. Finally, conclusions are drawn in Section 6.

QUALITY ORIENTED RQR CHAIN

RQR Chain

In the manufacturing process, normal and abnormal variations from man, machine, material, method, measurement, and environment (5M1E) should be accumulated and inherited by the work pieces, resulting in the variations of product dimensions eventually. Variations of product quality formed in the manufacturing process are the basic factors influencing the product inherent reliability. As manufacturing system acts as the material carrier of manufacturing process, its stability determines the quality of the manufactured product. Thus, it can be concluded to a great extent that the product inherent reliability is relying on the reliability of manufacturing system with a fixed product design scheme, product reliability in manufacturing process would be deteriorated by those abnormal factors like wear degradation or failures of system components or some assembly errors, and so forth, and the number of potential quality defects lying in the manufactured products should increase successively, resulting in the product inherent reliability failing to satisfy the design requirements. It is obvious that if the reliability of manufacturing system cannot be guaranteed effectively, the dimensional parameters of the manufactured work pieces will be deteriorated constantly (such as abnormal dimensional variations in the manufacturing process), and contrasted with design reliability, these variations would bring about a decline of product inherent reliability. To make matters worse, products composed by these unreliable components are prone to some unpredictable fatal failures. In order to describe the relationship among manufacturing system reliability Rm, process quality Q_p, and product inherent reliability R_p, a conceptual model of RQR chain is put forward and shown in Figure 1.

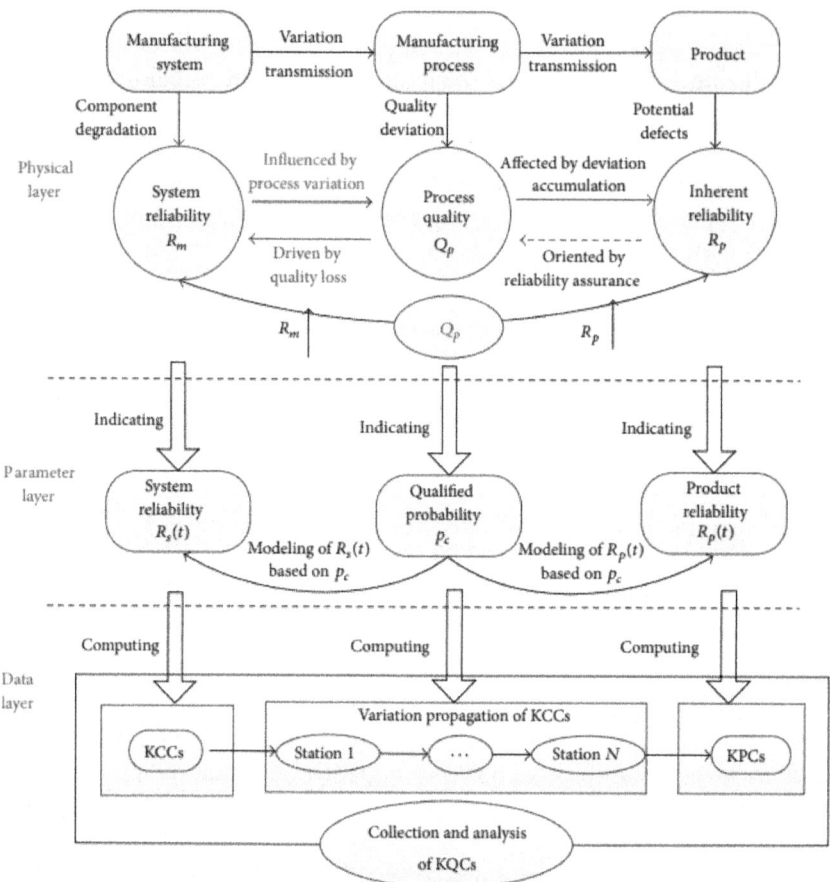

Figure 1: Conceptual model of RQR chain.

As shown in Figure 1, the product quality in manufacturing process is the linkage of the manufacturing system reliability and product inherent reliability which is clearly reflected by the RQR chain, and in the view of the mathematical modeling, the RQR chain could be divided into the following three layers from the top-down: physical layer, parameter layer, and data layer. The three-tier construction of RQR chain could enable the integrated analysis for the reliability of manufacturing system based on the quality data in manufacturing and reliability data in field as much as possible.

(1) Physical Layer of RQR Chain. The coeffects of manufacturing system, manufacturing process, and product are described clearly in this layer. Specifically, due to system components wear or failures caused by friction wear, reliability of manufacturing system Rm gets lower. During the

execution of functional requirements from manufacturing system, quality in manufacturing process Q_p results in abnormal variations accompanied by degradation of the manufacturing system. Correspondingly, owing to these quality variations, there may be potential defects retained in the final product, which has a negative effect on product reliability R_p. That is to say, reliability of manufacturing system determines directly the stability of manufacturing process quality, and then the reliability of produced product is subjected to the stability of manufacturing process quality.

(2)   Parameter Layer of RQR Chain. The indicating parameters of the mentioned coeffects in chain are given in this layer, the quantitative indicators of manufacturing system reliability, product qualified probability, and product inherent reliability are presented, and these parameters are the carriers of the coeffects in chain.

(3)   Data Layer of RQR Chain. The product key quality characteristics including key control characteristics and key product characteristics (KQCs) are presented in the data layer, which provides the data source to compute the value of parameters given in the parameter layer

As shown in Figure 1, the quality of manufacturing process Q_p is the core of the RQR chain, which links manufacturing system reliability Rm and product reliability R_p. To accurately analyze manufacturing system reliability oriented by product inherent reliability and further optimize the manufacturing system, quantifying the quality of manufacturing process is bound to the prerequisite. Typically, current studies adopt product qualified rate to characterize the quality of manufacturing process, which simply reflects the cursory state of process quality Q_p in the form of scalar quantity. And it is still actually inconvenient to carry out system reliability Rm modeling and optimization that considers process quality variation information in the form of vector. Therefore, based on the definition of product quality, the big data like variation vector of manufacturing process and system components degradation is combined firstly to put forward product qualified probability to represent manufacturing process quality in vector

Product Qualified Probability

Product qualified rate is often used in traditional quality control to describe the quality of manufactured products and the capability of manufacturing process. It is deemed that if all the key quality characteristics of products are within their tolerance limits at time t, products are thought to pass through the inspection and there is no substandard product throughout the manufacturing process. Count the number of qualified products and figure out how much it accounts for the total inspected products, and, namely, the product qualified rate is obtained. Here, product qualified rate integrates count information of a single product as a whole and characterizes the quality of manufactured products in the form of scalar. However, the values of the detected quality characteristics are obtained in the vector form with the advancement of detection technology in practical engineering applications usually. Different from the scalar expression of product qualified rate, defining quality levels and grades often requires quantifying the extent of how values of product key quality characteristics approximate the predetermined targets further. And thus, measurable information such as parameters of different components inside a single product should be the concern of quality control. At present the qualified rate is basically the only evidence to determine whether a batch product is qualified or not, and the judgment is arbitrary, which will result in neglecting and omitting useful variation information of those measurable key quality characteristics under modern vectorial measurement environment. And it turns out to be not conducive to fully exploit and utilize process quality data and then carry out a comprehensive analysis of both product reliability and manufacturing system reliability. Accordingly, product qualified probability oriented by quality grade is proposed and the vector space of different parameters that constitute product key quality characteristics is built in this paper, and thus vectorization of qualified degree of products is realized. When values of the key quality characteristics are closer to the predetermined ones, both quality grade and the corresponding degree of qualified product become higher. Namely, it corresponds to a higher product qualified probability. From the perspective of population and sample, Figure 2 contrasts the differences and relations between traditional qualified rate and the proposed qualified probability based on the interactive scalar of the whole population and the vector of the single sample.

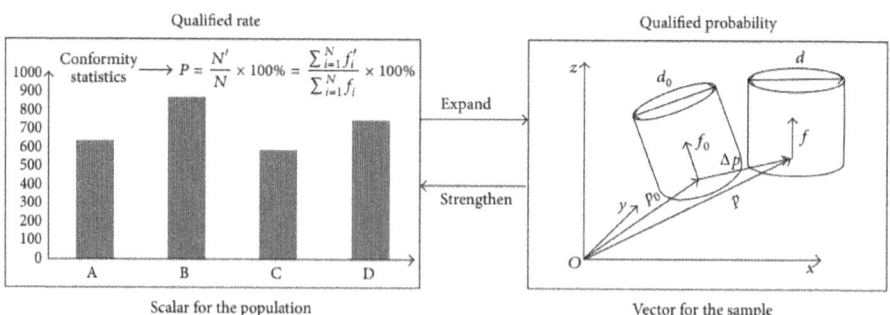

Figure 2: Comparison of qualified rate and qualified probability.

Given the threshold value θ_k of key quality characteristics, product qualified probability can be expressed as

$$P_k^c = \Pr\left\{\Delta y\left(k\right) \le \theta_k\right\}. \tag{1}$$

Here, $\Delta y(k)$ denotes the variation of the actual characteristic dimension to the target value at station k, and it can be obtained through the SoV [23] (Stream of Variation) theory.

Generally, the threshold value θ_k can be acquired by means of product quality specifications and the popular process capability index. Set the tolerance range of products to be USL (upper specification limit) and LSL (lower specification limit), with process capability index being C_{pk} (actual process capability) and variance of key quality characteristics being θ_v, and their values should be calculated via the formula of $\ddot{C}_{pk} = (\text{USL} - \text{LSL})/6\sqrt{\theta_k}$, $\theta_k = [6C_{pk}/(\text{USL} - \text{LSL})]^2$.

Set $f(x) = \Delta y(k) - \theta_k$, and ignore the noise of manufacturing process and the measurement. Since (x) is a nonlinear function of $x(k) = [x_1, x_2,...,x_n]^T$, for simplicity, Taylor expansion is used to linearize $f(x)$ as below:

$$f\left(x\right) = a_0 + \sum_{i=1}^{n} a_i x_i \quad \left(a_i \text{ is constant}\right),$$

$$\beta = \frac{\mu_f}{\sigma_f} \tag{2}$$

$$= -\frac{a_0 + \sum_{i=1}^{n} a_i \mu_{x_i}}{\sqrt{\sum_{i=1}^{n} a_i^2 \sigma_i^2 + \sum_{i=1}^{n} \sum_{\substack{j=1 \\ j \ne i}}^{n} a_i a_i \text{Cov}\left(x_i, x_j\right)}}$$

$$\left(\mu_f = a_0 + \sum_{i=1}^{n} a_i \mu_{x_i}\right). \tag{3}$$

As $\mu_x = \left(\mu_{x_1}, \mu_{x_2}, \cdots, \mu_{x_n}\right)$ is the mean point of independent variables of , correspondingly, mean of the dependent variable is denoted by $\mu_f = f\left(\mu_{x_1}, \mu_{x_2}, \cdots, \mu_{x_n}\right)$ and the standard variation by $\sigma_f^2 = \sum_{i=1}^{n} a_i^2 \sigma_i^2 + \sum_{i=1}^{n} \sum_{\substack{j=1 \\ j \ne i}}^{n} a_i a_i \text{Cov}(x_i, x_j)$. When process variables obey iid (independent and identically distributed), there exists $\sigma_f^2 = \sum_{i=1}^{n} a_i^2 x_i^2$. And product qualified probability of manufacturing process should be derived as

$$P_k^c = \Pr\{\Delta y(k) \le \theta_k\} = \Pr\{f(x) \le 0\}$$

$$= p\left\{\frac{f(x) - \mu_f}{\sigma_f} < -\frac{\mu_f}{\sigma_f}\right\} = \Phi(-\beta).$$

$$(4)$$

RELIABILITY MODELING OF MANUFACTURING SYSTEM BASED ON RQR CHAIN

Analysis of Product Inherent Reliability Oriented by Qualified Probability

Manufacturing processes are comprised of raw materials purchasing, parts machining, components assembling, and performance testing. With reference to reliability requirements of the design phase, parts that are either outsourced or homemade are operated by machining, assembling, testing, and so forth, and thus the Work in Process (WIP) or the end products are finally produced. Influenced by a variety of variation factors, the product inherent reliability gets lower than requirements of design reliability usually. Jiang and Murthy [4] noted that nonconforming components and assembly errors would have undesirable effect on the manufactured product reliability, and the nonconforming components are the basic adverse factors. Based on this standpoint, the product qualified probability is brought forward to measure the potential influence of nonconforming components on the product inherent reliability.

Assuming that failure distribution of products in the design phase is $F0(t)$, the design reliability of products $R_o(t)$ can be expressed as $R_o(t) = 1 - F_o(t)$. Correspondingly, failure density function and failure rate function are denoted by $f_o(t) = dF_o(t)/dt$ and $r_o(t) = f_o(t)/R_o(t)$, respectively. Similarly in the manufacturing process, failure distribution of nonconforming components is $H(t)$ with reliability function being $R_h(t) = 1 - H(t)$, failure density function being $h(t) = dH(t)/dt$, and failure rate function being $r_h(t) = h(t)/R_h(t)$. According to the number of the nonconforming components and the unqualified degree of the nonconforming components, define the probability that results in nonconforming components as q and the qualified probability of components as p_c.

Then with all the components being fully qualified, namely, $p_c = 1$, it is the only case of occurrence of nonconforming components that affects the reliability of manufactured products, which is denoted by

$$R_1(t) = (1 - q) R_o(t) + qR_h(t).$$ (5)

Obviously, with q being equivalent to 0, the manufactured reliability equals the design reliability, indicating that no nonconforming components occur, and with q being equivalent to 1, it happens to be the opposite from the former case of $q=0$, indicating that all components fail to be conforming in manufacturing process.

When it comes to components being not fully qualified with $0<pc < 1$, the manufactured reliability is subjected to the occurrence of nonconforming components and the related unqualified level jointly, and, then, inherent reliability of the manufactured product is as follows:

$$R_2(t) = (1 - q)(1 - p_c) R_o(t) + qp_cR_h(t).$$ (6)

It is evident that with $p_c = 1$, $R_2(t) = qR_h(t) \leq$, $R_1(t) = (1-q)R_o(t)+qR_h(t),$, which shows that integration of both the occurrence of nonconforming components and their unqualified level is conducive to a more precise estimation of the manufactured reliability. While simply based on the case whether components are unqualified, inherent reliability of the manufactured products will be blindly overestimated, which would have a great negative impact on the objective analysis and optimization for the reliability for manufacturing system.

Modeling Reliability of Manufacturing System Based on Qualified Probability

Reliability of the manufacturing system is an important factor to ensure product quality and productivity. Considering the difficulty in quantifying the reliability issues timely caused by product dimension variations, usually those reliability issues are easily neglected in system design and optimization. And unfortunately, with many failures or performance degradation failing to be diagnosed in real time, further identification or predication is of nonsense, which may result in unnecessary downtime for machines and reduce the production efficiency. Mechanisms of performance degradation and occurrence of failures of system components are versatile and complicated. Both components themselves of this current station and quality variations from the upstream station make a difference to the performance of components as well as the manufacturing system. Reliability analysis of multistation manufacturing system notes that via the transmission of quality variations, interaction between variations of key product characteristics (KPCs) and reliability of system components is also transferred through the stations. Thus, the statistical correlation between manufacturing system component failures and unqualified products could be shown in Figure 3.

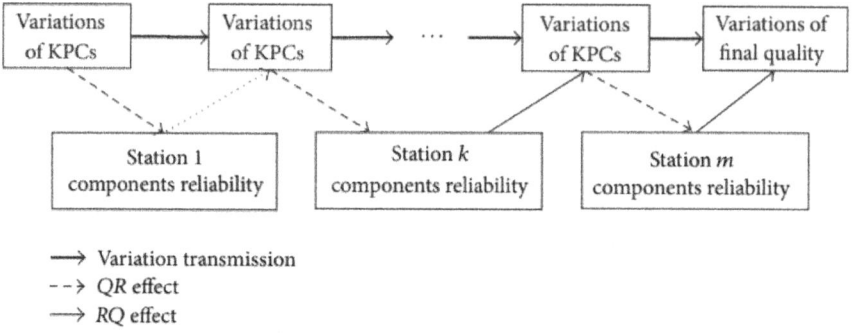

- $\longrightarrow$ Variation transmission
- $-->$ QR effect
- $\longrightarrow$ RQ effect

Figure 3: Interaction between quality variations and components reliability in multi-station process.

Accordingly, we are badly in need for a new connotation of reliability of manufacturing system. Degree of reliability is generally used to measure the reliability of the manufacturing system which refers to the probability a system completes its intended functions under specified conditions within the specified time. Not only the expected functions of a manufacturing process should consider the uptime of manufacturing system, but also the quality of manufactured product should also be involved to holistically assess the performance and reliability of manufacturing system. As a conclusion, the reliability of manufacturing system R(t) should be identified as the probability that system components do not fail themselves and at the same time manufactured products are completely qualified within a period of time, which can be expressed as

$$R(t) = P\{\text{system does not fail by time } t$$

$$\cap \text{ product quality is qualified by time } t\}. \tag{7}$$

Define R^F and R^Q to represent cases where system does not suffer catastrophic failures and product quality is qualified by time t, respectively. And $_{(tk)}$ means the performance state of system components at the endpoint of t_k. To sum up, the reliability of manufacturing system could be rewritten as

$R(t) = P\{\text{System does not fail by time } t$

$\cap \text{ Product quality is qualified by time } t\} = \Pr\{R^F$

$\cap R^Q\} = \Pr\{R^F \mid Z(t)\} \cdot \Pr\{R^Q \mid Z(t)\}.$

$$(8)$$

Here, $\Pr\{R^Q \mid Z(t)\}$ corresponds to the product qualified probability mentioned in Section 2.2. Namely, there exists $\Pr\{R^Q \mid Z(t)\} = P_k^c = \Phi(-\beta)$. So,. So, the process of how to determine the $\Pr\{R^F \mid Z(t)\}$ will be highlighted in the following.

In general, the performance state and relative operating conditions of system components determine the reliability of manufacturing system. Either components failures or degradation by wear could cause a decline in reliability of manufacturing system. Since status information like wear or degradation of tools comes along the running of manufacturing system, failure data may not occur necessarily with the advancement of manufacturing technologies. Therefore, this paper prefers to consider information of wear or degradation of system components as the main factors affecting the reliability of manufacturing system. The wear and tear of system components are accumulated and increased along with the front and back work stations one by one. Let $\Delta(k)$ be the amount of wear from the individual station k, and the cumulative amount of wear by station k is $Z(k) = Z(k-1) + \Delta(k)$. If all the wear or degradation processes are independent and identically distributed as most mechanical products, according to the central limit theorem, (k) can be rewritten as

$$Z(k) = \sum_{j=1}^{k} \Delta(j) \approx N(k \cdot E[\Delta(j)], k \cdot \text{Var}[\Delta(j)]).$$

$$(9)$$

Define $\lambda(t)$ as the probability that system component fails at station $k+1$ while it still functions well at station k by time t, namely, the failure rate of the system component. Assuming that failure of the individual component is subjected to an exponential distribution (for high reliability of complex systems, exponential distribution can approximately model the failure distribution for those system components), reliability of system components at station k is expressed as (excluding the impact of component wear on component strength, the failure rate can be regarded as a constant):

$$R_k^F(t) = e^{-\lambda_k(t) \cdot t}.$$

$$(10)$$

When the input products have problems of quality variations, wear and tear of system components will be accelerated as

$$\lambda_k(t) = \lambda_0(t) + E\left(\alpha_k\left(X(k) - m_k\right)^2\right).$$
(11)

Here, $\lambda_0(t)$ is the initial failure rate irrespective of impact of input product quality on system components; $X(k)$ is the practical quality index at station k; mk stands for the standard value of $X_{(k)}$ with α_k being the correction coefficient which reflects the impact of input quality variation on wear of components. And the reliability of the whole system could be presented as

$$R^F(t) = \prod_{i=1}^{n} R_i^F(t).$$
(12)

Suppose that η is the allowable maximum amount of components wear, ε_k is the degradation coefficient of relevant performance, w_0 is the initial rate of degradation which corresponds to time $t_k = 0$, and the probability that no failures occurred for the whole system by time t is as follows:

$$\Pr\left\{R^F \mid Z(t)\right\} = P\left(\prod_{i=1}^{n} R_i^F \mid Z(t) < \eta\right)$$

$$= P\left(\prod_{i=1}^{n} e^{-[\lambda_{0i}(t) + E(\alpha_i(X(i) - m_i)^2)] \cdot t_i} \mid \sum_{k=0}^{h} w(t) < \eta\right)$$

$$= P\left(\prod_{i=1}^{n} e^{-[\lambda_{0i}(t) + E(\alpha_i(X(i) - m_i)^2)] \cdot t_i} \mid h\left(w_0 + e^{-\varepsilon_k \cdot t}\right)\right.$$

$$\left. < \eta\right)$$

$$= \prod_{i=1}^{n} \exp\left(-\left[\lambda_{0i}(t) + E\left[\alpha_i\left(X(i) - m_i\right)^2\right]\right] \cdot t_i\right)$$

$$\cdot \int_0^{\eta} \exp\left(-h\left(w_0 + e^{-\varepsilon_k \cdot t}\right)\right) d\varepsilon.$$
(13)

Combined with formula (8), the final expression of the reliability of manufacturing system based on the product qualified probability (t) could be written as

$$R(t) = \Pr\left\{R^F \mid Z(t)\right\} \times \Pr\left\{R^Q \mid Z(t)\right\}$$

$$= \exp\left(\sum_{i=1}^{n}\left(-\left[\lambda_0(t) + E\left[\alpha_i\left(X(i) - m_i\right)^2\right]\right] \cdot t\right)\right)$$

$$\cdot \int_0^{\eta} \exp\left(-h\left(w_0 + e^{-\varepsilon_k \cdot t}\right)\right) d\varepsilon_k$$

$$\cdot \Phi\left(-\frac{\alpha_0 + \sum_{i=1}^{n}\alpha_i\mu_{x_i}}{\sqrt{\sum_{i=1}^{n}\alpha_i^2 + \sum_{i=1}^{n}\sum_{\substack{j=1\\j\neq i}}^{n}\alpha_i\alpha_j\mathrm{Cov}\left(x_i, x_j\right)}}\right)$$

$$= \exp\left(\sum_{i=1}^{n}\left(-\left[\lambda_0(t) + E\left[\alpha_i\left(X(i) - m_i\right)^2\right]\right] \cdot t\right)\right)$$

$$\cdot \int_0^{\eta} \exp\left(-h\left(w_0 + e^{-\varepsilon_k \cdot t}\right)\right) d\varepsilon_k \Phi\left(-\beta\right).$$

$$(14)$$

According to formula (14), if the interaction between product quality and components failures or performance degradation has been ignored, the reliability of manufacturing system should be overestimated, which would endanger the quality and reliability of manufactured products, and the final goal to obtain high reliable products cannot be fulfilled. The correlation modeling of product quality and components reliability should help us to establish a more objective and accurate model, which makes a more authentic and practical assessment of the reliability of manufacturing system, and provides specific goals and directions for further improving the reliability of manufacturing system.

PREVENTIVE OPTIMIZATION STRATEGY FOR RELIABILITY OF MANUFACTURING SYSTEM DRIVEN BY QUALITY LOSS IN MANUFACTURING PROCESS

Requirements Analysis of Dynamic Optimization

The purposes of conducting analysis, modeling, and assessment of the reliability manufacturing system are mainly designed to guide the timely maintenance for failed components or degraded ones, which would help to improve the quality level of the manufactured products, and how to obtain the optimal strategy in real time of the reliability of manufacturing system has long been a mathematical puzzle in industrial engineering [16, 17].

The previously proposed reliability model of manufacturing system highlights the role of product quality variation and its interaction effect with components reliability. Therefore, not only maintenance cost of manufacturing system itself but also the quality loss in manufacturing process should be simultaneously considered when developing reliability optimizing strategies for manufacturing system. That is to say, the total costs for optimizing manufacturing system should include quality loss caused by the variations that occurred in manufacturing process, maintenance costs by component failures, and so forth. According to the reliability model presented in Section 3.2, in order to realize dynamic or real time optimization, the product quality loss should be taken into account to formulate the optimization strategies. For the convenience of the computation, the assumptions and hypotheses are given firstly as below:

(1)   Suppose the maintenance cycle is T, and $T = k\Delta t$, $k \in N$, where k is a constant. Then the extent of performance degradation for each component can be determined. Denote the degradation state of component i by z_i ($i = 1, 2, \ldots, n$) and assume that the cost needed for conducting a state inspection for component i is C_i, where the inspection time can be neglected. Without loss of generality, Δt can be considered as 1, and thus T can be directly written as k in the context. Since the operating time for each component may be different from each other, time for state inspection separately may be not necessarily synchronized.

(2)   Assume there are two ways of maintenance of each component, which includes the postmaintenance and preventive maintenance, and inspection time of the maintenance is fixed at t_k. If the maintenance is assumed to be an overhaul or a complete replacement, functions of components are believed to be fully restored by repairment. Define costs for postmaintenance and preventive maintenance as C_c^i and C_p^i, respectively, where the magnitude relation conforms to $C_c^i > C_p^i$.

Optimization Strategy Decision-Making Model Driven by Quality Loss

Based on the above analysis and assumptions, the objective of optimization strategy for the reliability of manufacturing system is to minimize the costs of quality loss, postmaintenance, and preventive maintenance simultaneously. And it can be represented as a constraint optimization problem in the following expression:

$$\text{Min EC} = \text{Min } f\left(C_m, C_q\right), \tag{15}$$

where E_C represents the expected cost of the optimization strategy, C_m means the expectation of average maintenance costs comprised of postmaintenance costs and preventive maintenance costs, and C_q signifies the expectation of average quality loss.

It is generally believed that failure rate of each system component is fixed to the same as λ_i. And to be consistent with the prementioned assumption in Section 3.2, the failure of the individual component is subjected to an exponential distribution, and the reliability of system components at station i is expressed as $e^{-\lambda_i \cdot t}$. Accordingly, the preventive maintenance at the unit expense of C_p^i for the reliable system component gets the cost of $C_p^i e^{-\lambda_i t}$, whereas the postcorrective maintenance at the unit expense of C_c^i for the unreliable system component gets the cost of $C_c^i(1 - e^{-\lambda_i t})$. And thus, for one single system component by each inspection time, the total maintenance cost generally consists of the two mentioned parts of the preventive $C_p^i e^{-\lambda_i t}$ and the postcorrective $C_c^i(1 - e^{-\lambda_i t})$.

What is more, with lifetime of component i being s_i and the number of these system components being n, the average maintenance costs C_m can be extended as

$$C_m = \sum_{i=1}^{n} \frac{C_p^i e^{-\lambda_i t} + C_c^i \left(1 - e^{-\lambda_i t}\right)}{\int_0^{s_i} e^{-\lambda_i t} dt}.$$
(16)

And it can be simplified as the following expression:

$$C_m = \sum_{i=1}^{n} \lambda_i \left[\frac{C_p^i}{1 - e^{-\lambda_i s_i}} + C_p^i - C_c^i \right].$$
(17)

And the quality loss function based on the Taguchi function is expressed as

$$L(k) = q\left(X(k) - m_k\right)^2.$$
(18)

Here, (k) is the quality loss with q the constant responding to the coefficient of quality loss, $X(k)$ representing product key quality characteristic, and m_k being the specification value for $X(k)$.

Take the mathematical expectation for formula (18) as

$$E[L(k)] = q\mathrm{Var}[X(k)] + q\left[E(X(k)) - m_k\right]^2,$$
(19)

wherein $E[X(k)]$ and $\mathrm{Var}[X(k)]$ are the mathematical expectation and variance of $X(k)$, respectively.

When quality loss is taken into account of the optimization costs, estimation for the costs will be more conservative actually. After applying the exponential distribution to describe cost of quality loss by time t, the expectation of average quality loss C_q should be rewritten as

$$C_q(t) = C_0 \exp\{-E[L(k)] \cdot t\},$$

(20)

where C_0 is the initial cost, meaning there is no loss of product quality

Integrating formula (16) and formula (20) into formula (15), the optimization objective is converted to minimize both maintenance costs and product quality loss, and the final formula is as follows:

$$\text{Min EC} = \text{Min } f\left(\sum_{i=1}^{n} \lambda_i \left[\frac{C_p^i}{1 - e^{-\lambda_i s_i}} + C_p^i - C_m^i\right],\right.$$

$$\left.\lim_{T \to \infty} \frac{\int_0^T C_0 \exp\{-E[L(k)] \cdot t\} \, dt}{T}\right).$$

(21)

CASE STUDY

Background

As a key part of the engine, the cylinder head is mounted on the upper end of the cylinder block with cylinder head bolts, forming a sealed combustion chamber together with the cylinder block. Coordinating with components of intake and exhaust valves, fuel injectors, pneumatic valves and others, the cylinder head plays a vital role in controlling fully combustion of the air and fuel inside the cylinder. And thus, the system and process for manufacturing the cylinder head turns to be particularly complex and elaborated, key quality characteristics like the surface roughness, geometrical shape, machining dimensions and location precision are needed special attention and monitoring. Usually, variations of dimensions have a great impact on the assembly precision and even the overall performance of the engine. The cylinder head is comprised of the following six components: a superface, a subface, a frontface, a backface, a side-entry face, and a side-outlet face. The machining features are reflected in the complicated structures of surfaces and holes. Accordingly, how to assure the machined surfaces and holes with high precision is the core function of the manufacturing system of cylinder heads, which includes function modules of cutting, clamping, controlling, testing, and clearing. The key operation processes of the cylinder head manufacturing system are shown in Figure 4.

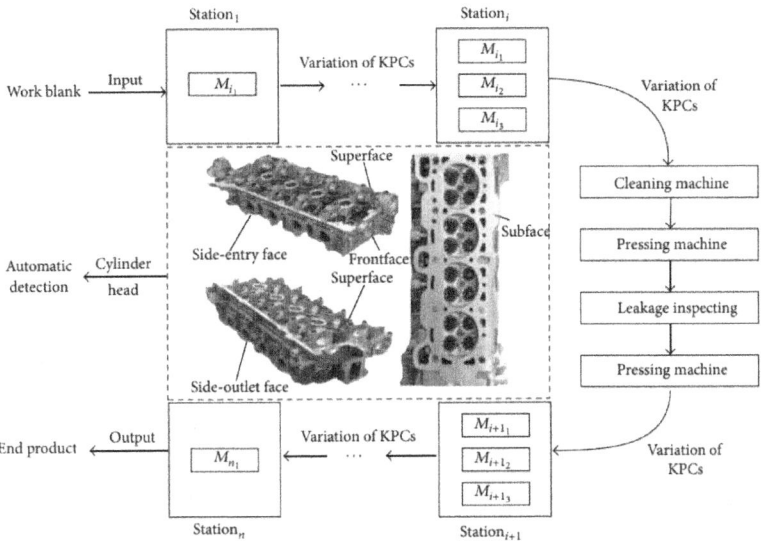

Figure 4: An illustration of key manufacturing process for cylinder head manufacturing system.

In this paper, key design characteristics of the cylinder head machined by the studied manufacturing system are listed in Table 1.

Table 1: Key design characteristics of the cylinder head

Basic dimension	Machining equipment	Process description	Major characteristics	Design specification
B_1	Machining center	Drilling hole A	Diameter	$\phi 13^{+0.018}_{0}$
B_2	Machining center	Drilling hole B	Diameter	$\phi 12.5^{+0.018}_{0}$
B_3	Horizontal machine center	Milling face C	Distance	121.3 ± 0.1
B_4	Horizontal machine center	Hinging hole D	Diameter	$\phi 10^{+0.15}_{0}$

In practical manufacturing process, due to the integrated effects of various variations, reliability of the cylinder head fails to meet the designed reliability requirements, resulting in a phenomenon of reliability degradation relative to the design reliability and an unsatisfactory response from customers. With mechanical inspection highly automated, it is urgent to reduce the interference from the vector space of key product characteristics and further to monitor the cylinder head quality effectively throughout the entire manufacturing process. Typically, optimization of manufacturing system reliability is the fundamental premise to control the cylinder head quality and reliability. However, whether to replace worn tools or conduct regular maintenance for a high level of system reliability, the level of qualified rate and reliability of the produced products still remained low. How to conduct the reliability of manufactured product oriented modeling and optimizing of manufacturing system reliability are what the proposed RQR chain tries to contribute. With characteristics of B_1, B_2, B_3, B_4 identified as the key product characteristics, interaction of process product quality and system components reliability is established and the associated model of manufacturing system reliability and product quality oriented by product qualified probability is then created. Correspondingly, it is proved to play an important role in the reduction of dimension variations of the cylinder head, the decrease of tooling adjustments, the increase of manufacturing system reliability, and the drop in risk of degradation. Specific modeling processes are shown in Figure 5.

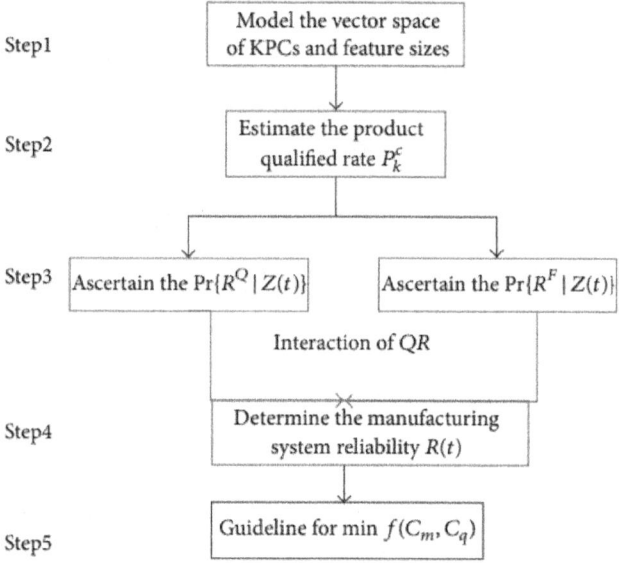

Figure 5: Flowchart of modeling and optimization.

Numerical Example

As shown in Figure 5, reliability analysis and the optimization example for the cylinder head manufacturing systems based on the proposed RQR chain are conducted in the following steps.

Create the vector space of key product characteristics as $X(p_{ix}, p_{iy}, p_{iz}, n_{ix}, n_{iy}, n_{iz})$ and ascertain the key feature dimensions of Yk. To be specific, in order to adapt the automatic detection process, the position vector pi and the orientation vector n_i ($i = 1, 2, 3, 4$) are defined, respectively, to quantify key characteristics of B_1, B_2, B_3, B_4 facing the three-dimensional space. Table 2 presents the parameters which the vector space $X(p_{ix}, p_{iy}, p_{iz}, n_{ix}, n_{iy}, n_{iz})$ ($i = 1, 2, 3, 4$) contains.

Based on the theory of SoV [23], transmission of key product dimensions between stations in manufacturing system is simplified as shown in Figure 6.

With the major variation u_k and noise of production wk (k = 1, 2, 3, 4) considered and regardless of noise of measurement γ_k, the produced product dimensions can then be expressed as

$$x(k) = A(k)x(k-1) + B(k)u_k,$$

$$y(k) = C(k)x(k).$$

(22)

Parameters of (k), $B(k)$, and $C(k)$ in formula (22) are ascertained by variation data of key product dimensions. With reference to what Table 2 has exhibited, use Mathematica software to calculate the four key dimensions as $(k) = [13.01, 12.496, 121.18, 10.07]^T$.

Table 2: Vector space model of key product characteristics

Number	Key characteristics	n_x	n_y	n_z	P_x	P_y	P_z
1	Hole A	0	1	0	41.5	22.5	0
2	Hole B	0	1	0	14.5.5	22.5	0
3	Face C	0	-1	0	91.5	-15.5	0
4	Hole D	1	0	0	0	0	61.5

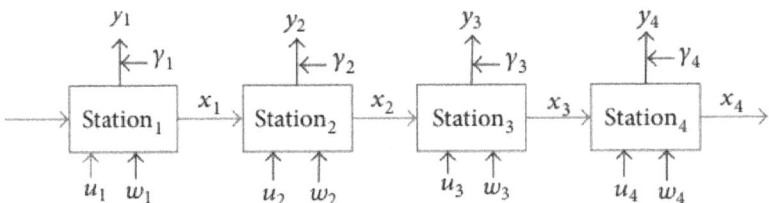

Figure 6: Simplified transmission model of key product dimensions.

Step 2. According to the obtained key dimensions Y_k, determine the product qualified probability P_k^c of manufacturing process and the impact $\Pr\{R^Q \mid Z(t)\}$ of process quality Q on reliability of manufacturing system. Assume that the observed original data follow a normal distribution, the related covariance

$$\sum_{i=1}^{4} \sum_{\substack{j=1 \\ j \neq i}}^{4} a_i a_j \text{Cov}(x_i, x_j) = 0$$

becomes

and μ_f and σ_f^2 are as follows:

$$\mu_f = f\left(\mu_{x_1}, \mu_{x_2}, \mu_{x_3}, \mu_{x_4}\right) = 0.382,$$

$$\sigma_f^2 = \sum_{i=1}^{4} a_i^2 \sigma_i^2 + \sum_{i=1}^{4}\sum_{\substack{j=1 \\ j \ne i}}^{4} a_i a_j \mathrm{Cov}\left(x_i, x_j\right) = 0.3454.$$

(23)

Referring to formula (3), the value of β should be computed as follows:

$$\beta = \frac{\mu_f}{\sigma_f} = \frac{0.382}{\sqrt{0.3454}} = 0.65.$$

(24)

And thus the product qualified probability is computed as

$$\Pr\left\{R^Q \mid Z(t)\right\} = \Phi\left(\beta\right) = \Phi\left(0.65\right) = 0.7422.$$

(25)

Step 3. Determine the impact of components reliability on manufacturing system reliability $\Pr\{R^F \mid Z(t)\}$ with process quality Q, wear loss, and failure rates of system components quantified, given that the fading rates $w(t_k) = w_0 + \exp(-\varepsilon_k \cdot t_k)$ are parameterized with $w_0 = 5 \times 10^{-5}$ and $\varepsilon_k = 1 \times 10^{-3}$. Meanwhile, the initial failure rate of the system component is $\lambda_{0i} = 6 \times 10^{-6}$ ($i = 1, 2, 3, 4$) and the interaction coefficients are $\alpha k = 3 \times 10^{-4}$, $h = 500$, and $\eta = 1.3 \times 10^{-2}$. Set the work time t for each station as 200 hours identically. With reference to formula (13), $\Pr\{R^F \mid Z(t)\}$ can be calculated as

$$\Pr\left\{R^F \mid Z(t)\right\}$$

$$= \exp\left(\sum_{i=1}^{n}\left(\left[\lambda_0(t) + E\left[\alpha_i\left(X(i) - m_i\right)^2\right]\right]\right) \cdot t_i\right)$$

$$\cdot \int_0^{\eta} \exp\left(-h\left(w_0 + e^{-\varepsilon_k \cdot t}\right)\right) d\varepsilon_k = 0.871.$$

(26)

Step 4. Based on the product qualified probability P_k^C and information of components reliability, the analysis model $R(t) = \Pr\{R^F \cap R^Q\}$ for manufacturing system reliability oriented by product inherent reliability is established. Integrate information of process quality $\Pr\{R^Q \mid Z(t)\}$ and components reliability into formula (14), and the reliability of manufacturing system (t) is finally estimated as

$$R(t) = \Pr\left\{R^F \mid Z(t)\right\} \times \Pr\left\{R^Q \mid Z(t)\right\}$$

$$= \exp\left(\sum_{i=1}^{n}\left(-\left[\lambda_0(t) + E\left[\alpha_i\left(X(i) - m_i\right)^2\right]\right] \cdot t\right)\right)$$

$$\cdot \int_0^{\eta} \exp\left(-h\left(w_0 + e^{-\varepsilon_k \cdot t}\right)\right) d\varepsilon_k \Phi(-\beta)$$

$$= 0.871 \times \Phi\left(\frac{0.382}{\sqrt{0.3454}}\right) = 0.871 \times 0.7422 = 0.646.$$

(27)

Step 5. Based on the above results, optimization strategy involved of process quality loss for manufacturing system is analyzed quantitatively. The intended idea is based on the comparison of the difference between the estimated R_{FQ} from the correlation model, namely, the $\Pr\{R^F \cap R^Q\}$, and the estimated RF irrespective of the correlation between process quality and components reliability, namely, the $\Pr\{R^F \mid Z(t)\}$.

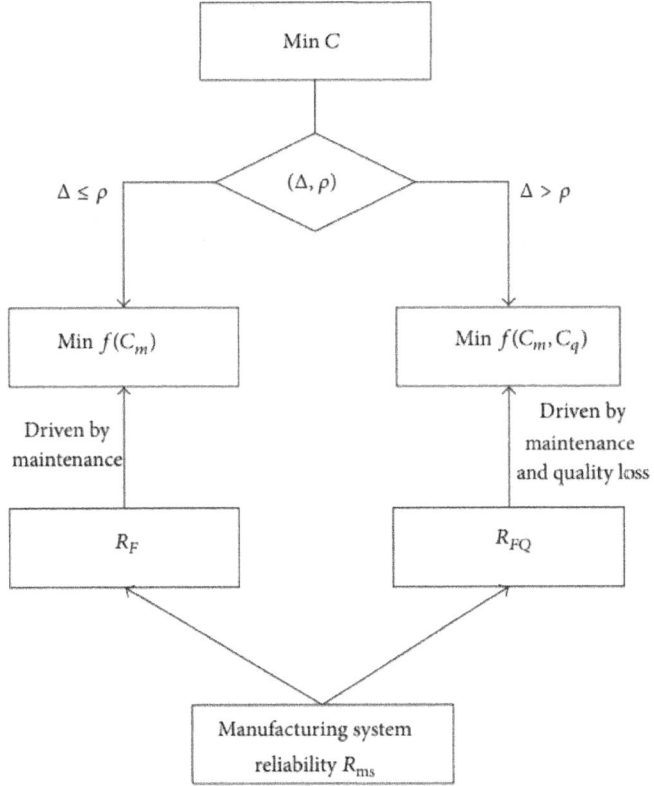

Figure 7: Optimization strategy for the manufacturing system.

With the difference $\Delta = R_F - R_{FQ}$ and the threshold value ρ ascertained, optimization strategy for manufacturing system is shown in Figure 7.

Result Analysis

From the results shown in Section 5.2, it is obvious that when the correlation between process quality and components reliability is not considered, the reliability of manufacturing system is approximately 0.871, whereas it turns to be 0.646 when the correlation is considered. This is to say, manufacturing system reliability is overestimated by 38.9%.

Choose time t as the independent variable; comparison between R_F and R_{FQ} is shown in Figure 8.

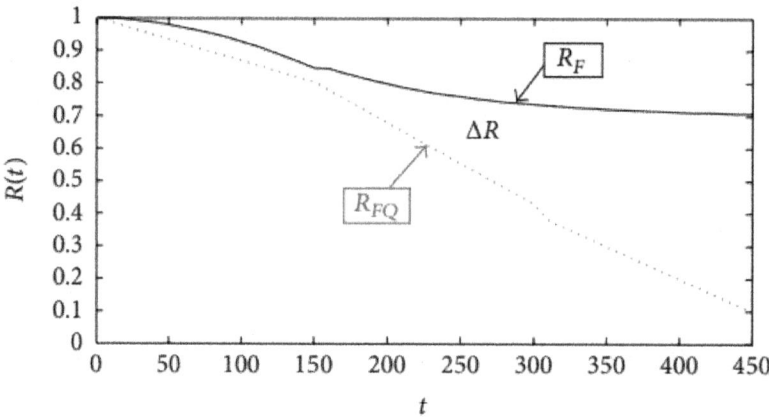

Figure 8: Comparison of manufacturing system reliability for the cylinder head.

As shown in Figure 8, the system reliability is often overestimated when we do not consider the interaction between product quality and components degradation or failures. Through the integral operation, the overestimated reliability is 44.97%. Additionally, the estimation error becomes more significant as time increases, which indicates that the cumulative transmission of quality variations does have a profound effect on the system reliability. Furthermore, in order to show the robustness of the proposed approach, sensitivity analysis of the reliability of manufacturing system on the product qualified probability is simulated, and the result is shown in Figure 9.

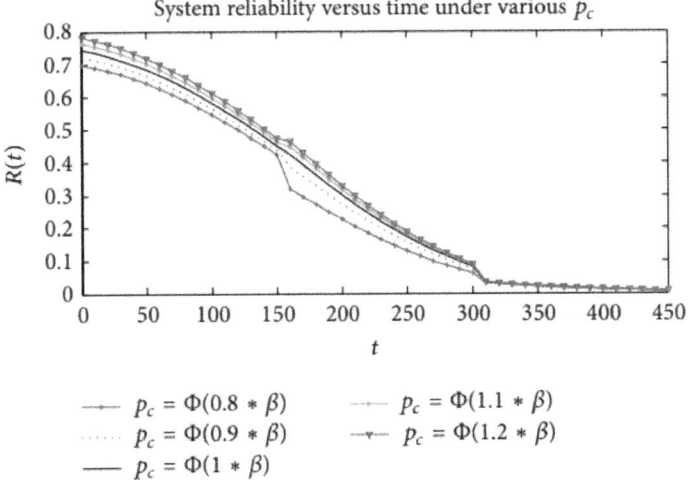

Figure 9: Sensitivity analysis of R on p_c for the cylinder head.

As shown in Figure 9, five levels of pc have been set to analyze the significant impact of the cylinder head qualified probability on the reliability of the manufacturing system (t). With a series of ±20% shift from the basic $\Pr\{R^Q \mid Z(t)\} = \Phi(\beta) = \Phi(0.65) = 0.7422$, namely, $p_c = \Phi(0.8 * \beta)$, $p_c = \Phi(0.9 * \beta)$, $p_c = \Phi(1 * \beta)$, $p_c = \Phi(1.1 * \beta)$, and $p_c = \Phi(1.2 * \beta)$, five reliabilitycurves of the manufacturing system are plotted versus time   under different levels of the product qualified probability to demonstrate the correlated sensitivity and robustness. And the synthetical simulation result shows that a high level of the proposed cylinder head qualified probability can usually guarantee a high level of the system reliability. Accordingly, the model of manufacturing system reliability based on the product qualified probability would help the reliability engineer to form a more objective and accurate mathematical model to analyze comprehensively the reliability and performance for manufacturing system. Moreover, the appropriate optimization strategies covering the maintenance for system components and adjustment of process scheme can be got, which should provide clear goals and directions to continuously improve the manufacturing system reliability.

CONCLUSION

In this paper, the RQR chain which could fuse the quality data and reliability data in manufacturing system reliability optimization is put forward, and the product qualified probability is introduced to quantify the impacts of quality variation in manufacturing process on the reliability of manufacturing system at the first time. Furthermore, a novel mathematical model of manufacturing system reliability is established based on the product qualified probability and RQR chain. Finally, the reliability and maintenance optimization strategy for manufacturing system is analyzed in view of total cost of maintenance costs and quality loss. The application result demonstrated that manufacturing system reliability tends to be overestimated if the mentioned interaction between product quality and manufacturing system reliability is omitted. The overrated value of the reliability of manufacturing system will not only deteriorate the produced product reliability and may also lead to a wrong maintenance strategy or miss the best opportunity for system maintenance.

To conclude, it is critically essential to consider the coeffects between process quality and system components reliability when modeling and assessing the reliability of manufacturing system. For future research, the following topics should be further expounded.(1)The improvement of the reliability optimization model based on the quality loss for different type of manufacturing system is needed. The coefficients of the optimization model are different for different manufacturing system; therefore, how to estimate accurately the coefficients from the big data from manufacturing system design, operation, and maintenance is planned.(2)The quantitative mathematical relationship of the manufacturing system reliability, manufacturing process quality, and the produced product reliability should be established successively. Specifically, the mathematical impact of the manufacturing system reliability on the produced product reliability should be constructed clearly, which should provide a solid foundation for the integrated reliability and maintenance optimization framework of the various types of manufacturing system.(3)In the last perspective research, we consider the aspect of reliability modeling and assessment in the design and setup of manufacturing system. The reliability level is determined in the design process of the manufacturing system, in order to satisfy the ever-increasing stringent quality and reliability requirements, the reliability design should be integrated with the functional design of manufacturing system, and new design theory like Axiomatic Design should be adopted into reliability design of manufacturing system.

ACKNOWLEDGMENT

This research was supported by Grant 61473017 from the National Natural Science Foundation of China.

REFERENCES

1.   Y. He, W. Linbo, Z. He, and M. Xie, "A fuzzy TOPSIS and rough set based approach for mechanism analysis of product infant failure," Engineering

Applications of Artificial Intelligence, 2015.

2. M. Colledani, T. Tolio, A. Fischer et al., "Design and management of manufacturing systems for production quality," CIRP Annals: Manufacturing Technology, vol. 63, no. 2, pp. 773–796, 2014.

3. P. Murthy, "New research in reliability, warranty and maintenance," in Proceedings of the 4th Asia-Pacific International Symposium on Advanced Reliability and Maintenance Modeling (APARM ‹10), pp. 504–515, McGraw-Hill International Enterprises, Wellington, New Zealand, December 2010.

4. R. Jiang and D. N. P. Murthy, "Impact of quality variations on product reliability," Reliability Engineering & System Safety, vol. 94, no. 2, pp. 490–496, 2009.

5. J. Li and N. Huang, "Quality evaluation in flexible manufacturing systems: a Markovian approach,"Mathematical Problems in Engineering, vol. 2007, Article ID 57128, 24 pages, 2007.·

6. J. Li and J. Lei, "Integration of manufacturing system design and quality management," IIE Transactions, vol. 45, no. 6, pp. 555–556, 2013.

7. R. R. Inman, D. E. Blumenfeld, N. Huang, J. Li, and J. Li, "Survey of recent advances on the interface between production system design and quality," IIE Transactions, vol. 45, no. 6, pp. 557–574, 2013.

8. N. Li and J. Ni, "Reliability estimation based on operational data of manufacturing systems," Quality and Reliability Engineering International, vol. 24, no. 7, pp. 843–854, 2008.

9. Y.-K. Lin and P.-C. Chang, "System reliability of a manufacturing network with reworking action and different failure rates," International Journal of Production Research, vol. 50, no. 23, pp. 6930–6944, 2012. ·

10. G.-D. Li, S. Masuda, D. Yamaguchi, and M. Nagai, "A new reliability prediction model in manufacturing systems," IEEE Transactions on Reliability, vol. 59, no. 1, pp. 170–177, 2010.

11. Y. Chen and J. Jin, "Quality-reliability chain modeling for system-reliability analysis of complex manufacturing process," IEEE Transactions on Reliability, vol. 54, no. 3, pp. 475–488, 2005.

12. Y. Chen and J. Jin, "Quality-oriented-maintenance for multiple interactive tooling components in discrete manufacturing processes," IEEE Transactions on Reliability, vol. 55, no. 1, pp. 123–134, 2006. ·

13. F. Zhang, J. Lu, Y. Yan, S. Tang, and C. Meng, "Dimensional quality oriented reliability modeling for complex manufacturing process," International Journal of Computational Intelligence Systems,

vol. 4, no. 6, pp. 1262–1268, 2011.

14. K. Rafiee, Q. Feng, and D. W. Coit, "Reliability modeling for dependent competing failure processes with changing degradation rate," IIE Transactions, vol. 46, no. 5, pp. 483–496, 2014.

15. N. Li, F. T. S. Chan, S. H. Chung, and A. H. Tai, "An EPQ model for deteriorating production system and items with rework," Mathematical Problems in Engineering, vol. 2015, Article ID 957970, 10 pages, 2015. ·

16. X. Gong, Y. Feng, H. Zheng, and J. Tan, "An adaptive maintenance model oriented to process environment of the manufacturing systems," Mathematical Problems in Engineering, vol. 2014, Article ID 537452, 10 pages, 2014.

17. L. Tlili, M. Radhoui, and A. Chelbi, "Condition-based maintenance strategy for production systems generating environmental damage," Mathematical Problems in Engineering, vol. 2015, Article ID 494162, 12 pages, 2015.

18. Z. Hajej, N. Rezg, and A. Gharbi, "Forecasting and maintenance problem under subcontracting constraint with transportation delay," International Journal of Production Research, vol. 52, no. 22, pp. 6695–6716, 2014.

19. H. Zied, D. Sofiene, and R. Nidhal, "Joint optimisation of maintenance and production policies with subcontracting and product returns," Journal of Intelligent Manufacturing, vol. 25, no. 3, pp. 589–602, 2014.

20. Z. Hajej, N. Rezg, and A. Gharbi, "A decision optimization model for leased manufacturing equipment with warranty under forecasting production/ maintenance problem," Mathematical Problems in Engineering, vol. 2015, Article ID 274530, 14 pages, 2015. ·

21. Z. Hajej, S. Turki, and N. Rezg, "Modelling and analysis for sequentially optimising production,maintenance and delivery activities taking into account product returns," International Journal of Production Research, vol. 53, no. 15, pp. 4694–4719, 2015.

22. L. Mifdal, Z. Hajej, and S. Dellagi, "Joint optimization approach of maintenance and production planning for a multiple-product manufacturing system," Mathematical Problems in Engineering, vol. 2015, Article ID 769723, 17 pages, 2015.

23. J. Shi, Stream of Variation Modeling and Analysis for Multistage Manufacturing Processes, CRC Press, Taylor & Francis, 2007.

Chapter 2

MODELING AND VERIFICATION OF RECONFIGURABLE AND ENERGY-EFFICIENT MANUFACTURING SYSTEMS

Jiafeng Zhang,[1,2] Mohamed Khalgui,[3] Wassim Mohamed Boussahel,[2,4] Georg Frey,[2] ChiTin Hon,[5] Naiqi Wu,[5] and Zhiwu Li[5,6]

[1]School of Electro-Mechanical Engineering, Xidian University, Xi'an 710071, China

[2]Chair of Automation and Energy Systems, Saarland University, 66123 Saarbrücken, Germany

[3]University of Carthage, 1054 Carthage, Tunisia

[4]Zentrum für Mechatronik und Automatisierungstechnik, 66121 Saarbrücken, Germany

[5]Institute of Systems Engineering, Macau University of Science and Technology, Taipa, Macau

[6]Faculty of Engineering, King Abdulaziz University, Jeddah 21589, Saudi Arabia

ABSTRACT

This paper deals with the formal modeling and verification of reconfigurable and energy-efficient manufacturing systems (REMSs) that are considered as reconfigurable discrete event control systems. A REMS not only allows global reconfigurations for switching the system from one configuration to another, but also allows local reconfigurations on components for saving energy when the system is in a particular configuration. In addition, the unreconfigured components of such a system should continue running during any reconfiguration. As a result, during a system reconfiguration, the system may have several possible paths and may fail to meet control requirements if concurrent reconfiguration events and normal events are not controlled. To guarantee the safety and correctness of such complex systems, formal verification is of great importance during a system design stage. This paper extends the formalism reconfigurable timed net condition/event systems

(R-TNCESs) in order to model all possible dynamic behavior in such systems. After that, the designed system based on extended R-TNCESs is verified with the help of a software tool SESA for functional, temporal, and energy-efficient properties. This paper is illustrated by an automatic assembly system.

INTRODUCTION

A reconfigurable manufacturing system (RMS) is designed at the outset for rapid change in structure, as well as in hardware and software components, in order to quickly adjust production capacity and functionality within a part family in response to sudden changes in market or in regulatory requirements [1]. A RMS should be designed with several configurations (behavior modes) to, respectively, meet different production requirements in various conditions. There are two types of reconfigurations: static and dynamic reconfigurations. Generally, a static reconfiguration is applied offline to modify a RMS extensively such as adjusting architecture of physical systems and removing obsoleted machines, whereas a dynamic system reconfiguration, to switch a RMS from one configuration to another at runtime, is applied with the aim of fault-tolerance or actively changing system behavior modes [2, 3]. This paper focuses on dynamic RMSs.

Traditionally, manufacturing is an energy-intensive process, using motors, steam, and compressed air systems to transform raw materials into durable goods and consumer products [4–6]. Recent research shows that switching machines of a manufacturing system into their energy-efficient modes when they are idle during production can make considerable contribution to the reduction of energy demand and thus can reduce carbon footprint as well as operating costs [7–15]. This paper takes the advantage of dynamic reconfigurations of machines of a RMS between their working modes and energy-efficient modes as a way of reducing system energy consumption. A RMS with such energy-efficient operations is called a reconfigurable and energy-efficient manufacturing system (REMS).

REMSs can be abstracted as reconfigurable discrete event systems (DESs) when only their logic behavior properties are investigated. In this paper, a reconfiguration is called a local reconfiguration, if it is applied for switching a machine of a REMS between its working mode and energy-efficient mode. A reconfiguration is named a global reconfiguration if it is applied for switching a REMS between different configurations.

A REMS should be able to reconfigure itself smoothly due to changed inner/outer environments at runtime. Meanwhile, normal unreconfigured events should go on occurring whenever they meet their occurrence preconditions.

However, uncontrolled concurrence of reconfiguration events and normal events may cause faults such as deadlocks and overflow [16–19]. Therefore, the formal verification is of great importance during design stages.

Petri nets [20, 21] have found an extensive application to discrete event systems [22, 23], including automated flexible manufacturing systems [24–28] and reconfigurable systems [29]. Reconfigurable timed net condition/ event systems (R-TNCESs) [30, 31] are reconfigurable extensions of timed net condition/event systems (TNCESs) [32, 33]. TNCESs [34, 35] have a visual graph expression, a clear modular structure, and an exact mathematical definition inherited from Petri nets [36–40]. In addition, they have a strong analysis software tool: SESA (http://homepages.engineering.auckland. ac.nz/vyatkin/tools/modelchekers.html) [41]. System behavior properties, such as state/event trajectories and temporal requirements, can be specified by Computation Tree Logic (CTL), extended CTL (eCTL), and timed CTL (TCTL) [42–44] before being checked by SESA automatically. If a property is satisfied by the system, the model checker will return "true". Otherwise, a counterexample will be returned. Therefore, TNCESs have been widely applied in verification and validation of industry control systems especially for manufacturing systems [45–47]. The verification of a R-TNCES can be performed with the assistance of SESA [30, 31].

However, R-TNCESs cannot fully meet our requirements for a REMS. In a R-TNCES, reconfiguration functions model system reconfiguration events and transitions model normal events. However, the concurrence of reconfiguration functions and transitions is forbidden in a R-TNCES, which is in fact inconsistent with system requirements of REMSs. As a result, formal verification of such complex systems cannot be performed.

Motivated by the fact aforementioned, this paper extends R-TNCESs. First, the reconfiguration functions of R-TNCESs are assigned with action ranges and concurrent decision functions. After that, they are divided into two types according to their action ranges: major and minor reconfiguration functions. The major ones are used to model global reconfiguration events, whereas the minor ones are applied to model local reconfiguration events. Accordingly, the dynamics of R-TNCESs is updated for these extensions such that the concurrence of reconfiguration events and normal events can be conditionally allowed to guarantee the system correctness. Afterwards, an implementation method for an extended R-TNCES is developed. Finally, the software tool SESA is applied to check system functional, temporal, and energy-efficient properties. An automatic assembly system is used to illustrate this work.

The paper is organized as follows. The system specification of REMSs and the applied automatic assembly system are depicted in Section 2. The drawbacks of R-TNCESs on analyzing REMSs and the proposed extended R-TNCESs are described in Section 3. The formal verification of a REMS based on extended R-TNCESs is illustrated in Section 4. Finally, Section 5 concludes this paper and briefly presents further studies.

RECONFIGURABLE AND ENERGY-EFFICIENT MANUFACTURING SYSTEMS

This paper treats a reconfigurable and energy-efficient manufacturing system (REMS) as a reconfigurable discrete event control system. This section presents system specification and interesting system dynamics before it illustrates them with an automatic assembly system.

System Specification

A REMS is designed with a set of configurations to meet various requirements in different execution environments. A configuration Con is defined as

$$Con = (Com, Str, Dat),$$

$$(1)$$

where Com is a set of all activated components in Con, Str defines the structure, that is, the connection relationship and the communication protocol among components of Com, and Dat denotes the set of all global variables and parameters of Con

A REMS is denoted by

$$Sys = \left(\sum, R_c \right),$$

$$(2)$$

where $\sum$ is the set of n configurations and $R_c : \sum \rightarrow \sum$ is the reconfigurable controller dealing with system reconfigurations.

There are two types of system reconfigurations in a REMS: global and local reconfigurations. The former ones are applied for switching system configurations. The latter ones are applied for switching an activated component between its working mode and energy-efficient mode when the system is in a particular configuration.

A REMS starts running as described in one of these configurations. After that, it should be able to change into other configurations smoothly due to the detection of component faults or other well-defined conditions. In addition, in each configuration, local reconfigurations can be applied to components such that the components can reconfigure themselves into their energy-efficient modes to save energy when they are idle and turn back to their working modes when the system needs them. Dynamics of a REMS can be described by the evolution of system states. The evolution is caused by the occurrences of events events. A REMS includes three types of normal events, local reconfiguration events, and global reconfiguration events.(1)If a normal event occurs, the system changes its state within its current configuration.(2) If a local reconfiguration event occurs, a component of current configuration switches into its energy-efficient mode or switches back into its working mode. (3)If a global reconfiguration event occurs, the system switches into another configuration.

Meanwhile, during a global or local reconfiguration, if normal events meet their occurring conditions and they are not modified by the occurring reconfiguration events, they should go on occurring. However, this kind of concurrence brings safety threat to the system, since they may cause unboundedness, deadlocks, and even other functional or temporal failings.

Running Example

An automatic assembly system, denoted by AAS, is applied to illustrate works presented in this paper. AAS includes three workstations ($W1$, $W2$, and $W3$) and four robots (Rb1, Rb2, Rb3, and Rb4). It is assumed that robots are high energy consumption machines. The respective time consumption of $W1$, $W2$, and W3 to finish a machining task is 40 time units, 30 time unites, and 50 time unites. The time consumption of both Rb1 and Rb2 to finish a task is 20 time units. The time consumption of both Rb3 and Rb4 to finish a task is 25 time units. The default working process diagram of AAS is shown in Figure 1.

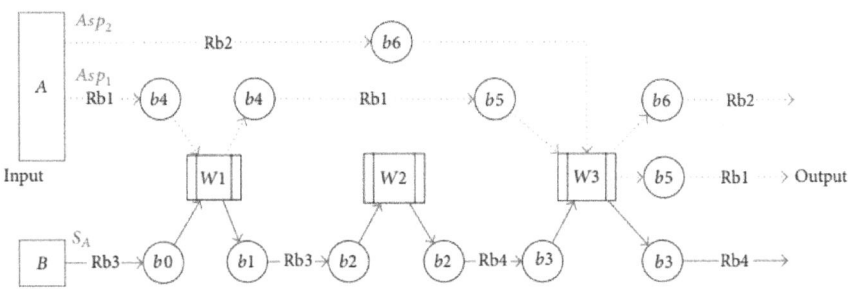

Figure 1: Default working process diagram of AAS.

The main function of AAS is to assemble machine parts into a subassembly of a vehicle, to be marked by S_A. Robots Rb1 and Rb2 move machine parts from the input into AAS, transfer machine parts between workstations, and remove trashy machine parts to the output. Dotted arrows in Figure 1 are used to denote the movements of machine parts during an assembly process. On the other hand, S_A is shifted along $W1$, $W2$, and $W3$ by robots Rb3 and Rb4. Solid arrows in Figure 1 are used to denote the movement of S_A. To make it clear, $b0$, $b1$,..., and $b6$ are used to denote positions where machine parts or S_A should be during an assembly process. The main assembly process is briefly described by the following three steps.

(1)   The to-be-worked subassembly SA is shifted from input B to $b0$ by Rb3. A machine part $Asp1$ is delivered to $b4$ from the input A. After that, Asp_1 and SA are preprocessed on W_1. Then, the preprocessed SA is moved to $b1$ automatically. The preprocessed Asp_1 is moved to $b4$ automatically before being moved to position $b5$ by Rb_1.

(2) S_A is transported to $b2$ from $b1$ by Rb3. Then, a second preprocess for S_A is done by $W2$. After that, S_A is shifted to $b3$ from $b2$ by Rb4.

(3)   A machine part Asp_2 is delivered to $b6$ by Rb2. Then, $W3$ starts the assembly after S_A is in $b3$, preprocessed Asp_1 is in $b5$, and $Asp2$ is in $b6$. After the assembly, the machined S_A is moved out by Rb4. Two other trashy machine parts are removed out of AAS by Rb1 and Rb2, respectively.

It is assumed that four behavior modes are designed for AAS. Their work processes are illustrated as follows.

(i):   *Mode* 1: *Mode* 1 is the default mode as depicted in Figure 1, where all the robots are applied.

(ii):   *Mode* 2: *Mode* 2 is a responding mode when Rb2 breaks down, where Rb1 should update itself to perform the function of Rb2

(iii):   *Mode* 3: *Mode* 3 is applied when Rb4 breaks down during the execution of *Mode* 1, where the work of Rb4 has to be done by Rb3.

(iv):   *Mode* 4: *Mode* 4 is applied when both Rb1 and Rb2 break down, where only Rb1 and Rb3 are applied. In this case, Rb1 should cover the function of Rb2 as in *Mode* 2 and Rb3 should cover the function of Rb4 as in *Mode* 3.

In each behavior mode, the applied robots should be able to reconfigure themselves into their energy-efficient modes when they are idle and reconfigure themselves back into their working modes when they have new tasks. A local reconfiguration for switching a robot from its working mode to its energy-efficient mode consumes one time unit. Likewise, a local reconfiguration for switching a robot from its energy-efficient mode back to its working mode consumes one time unit, as well.

To avoid the halt of a continuous production line, possible dynamic reconfigurations applied for switching AAS between these behavior modes are shown in Figure 2. The solid arrows denote global reconfigurations and dotted ones denote local reconfigurations.

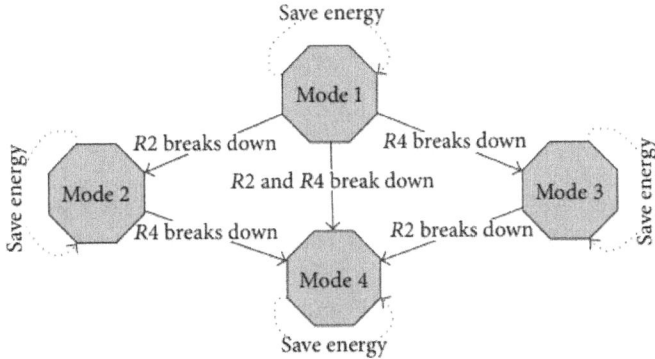

Figure 2: Possible reconfigurations in AAS.

It is assumed that a robot consumes one energy unit per time unit when it works in its working mode. However, it only consumes 30% energy units per time unit when it works in its energy-efficient mode. Note that the numerical value "30%" is an assumption by the authors to facilitate the quantitative analysis on energy-efficient operations. It does not come from any literature on industry systems.

Obviously, the possible reconfiguration events of AAS can occur simultaneously with many normal events in it. For example, when Rb1 is being modified by a global reconfiguration or being switched into its energy-efficient mode, only its own work needs to stop for a while and the workstations and other running robots should do their jobs unaffectedly.

Extended R-TNCES

Reconfigurable timed net condition/event systems (R-TNCESs) [30, 31] are extensions of timed net condition/event systems (TNCESs) [34, 35]. Reconfiguration functions of R-TNCESs can be used to model global reconfiguration events of REMSs. However, they are not proper to model local reconfiguration events of REMSs directly. In addition, the concurrence of normal events and reconfiguration events is currently not allowed in R-TNCESs. Therefore, in order to perform correct formal verification of a REMS, this paper extends R-TNCESs. This section briefly recalls basic conceptions of R-TNCESs, analyzes the drawbacks of R-TNCEs on investigating REMSs, and represents the proposed extended R-TNCESs.

R-TNCEs

Definition 1 (see [30]). A R-TNCES is a structure $RN = (B, R)$, where B is a behavior module and R is a control module. The behavior module B is a union of n superposed TNCESs. For any $i \in \{0,...,n-1\}$, the TNCES Γ_i is denoted by $\Gamma_i = (N_{\Gamma i}, z0_i)$ with $N_{\Gamma i} = (P_i, T_i, F_i, CN_i, EN_i, emi, DC_i)$. Then B can be represented as

$$\mathscr{B} = \left(\mathscr{P}, \mathscr{T}, \mathscr{F}, \mathscr{C N}, \mathscr{E N}, \mathscr{E M}, DC, \mathscr{L}_0 \right). \tag{3}$$

$P = P_0 \cup P_1 \cup \cdots \cup P_{n-1}$ (resp., $T = T_0 \cup T_1 \cup \cdots \cup T_{n-1}$) is a superset of places (resp., transitions). $F \subseteq (P \times T) \cup (T \times P)$ is a superset of flow arcs. $CN \subseteq (P \times T)$ (resp., $EN \subseteq (T \times T)$) is a superset of condition (resp., event) signals. $DC = DC_0 \cup DC_1 \cup \cdots \cup DC_{n-1}$ is a set of time intervals to input flow arcs. EM: $T \rightarrow \{-,,\}$ maps an event processing mode (AND or OR) for each transition. Let $Z_0 = (M_0, D_0)$, where $M_0 : P \rightarrow \{0, 1\}$ is the initial marking. $D_0 : P \rightarrow \{0\}$ is the initial clock position. $\Omega = \{N_{\Gamma 0}, N_{\Gamma 1}, ..., N\Gamma_{n-1}\}$ is a set of all TNCES structures that can be represented by RN.

The control module R is a set of reconfiguration functions. A reconfiguration function r is a structure $r = (Cond, s, x)$. Cond $\rightarrow \{true, false\}$ is the precondition of r. $s: \Omega \rightarrow \Omega$ is the structure modification instruction. $: R(N_{\Gamma i}, z_{0i}) \rightarrow Z_{0j}$ is the state correlation function, where Z_{0j} is a set of feasible initial states of Γ_j. $^*r = \Gamma_i = (N_{\Gamma}, z_{0i})$ (resp., $r^* = \Gamma_j = (N_{\Gamma j}, z_{0j})$) denotes the TNCES before (resp., after) the implementation of r. Definition 2 (see [30]). A state Z of R-TNCES RN is a pair $[N_i, z]$, where $N_{\Gamma} \in \Omega$ identifies the activated TNCES with $N_{\Gamma} = (P, T, F, CN, EN, em, DC)$, and $z = (m, d)$ is a state of N_{Γ} with $m: P \rightarrow \{0, 1\}$ and $d: P \rightarrow \{0, 1, 2, ...\}$.

In a R-TNCES, each TNCES in the behavior module models a configuration. For a R-TNCES RN, only one of the TNCESs of the behavior module B is activated at the beginning until a reconfiguration function is implemented. Other TNCESs with net structures defined in Ω can be activated only after implementing reconfiguration functions. At any time, only one of the TNCESs with net structures defined in Ω is activated.

If a reconfiguration function $r = (Cond, s, x)$ meets its precondition, that is, Cond = True, it is enabled. A reconfiguration function can fire if it is enabled, that is, to implement it. The evolution of a R-TNCES depends on what events (reconfiguration functions or transitions) take place. Let Γ_i be the activated TNCES with $\Gamma_i = (N_{\Gamma}, z_{0i})$, where $N_{\Gamma i} = (P_i, T_i, F_i, CN_i, EN_i, em_i, DC_i)$. If a maximal step $u \in_{Ti}$ fires, Γ_i evolves from its one inner state to another. However, if a reconfiguration function r fires, then Γi is transformed into Γ_j by changing its net structure and updating its state, where $^*r = \Gamma_i$, $r^* = \Gamma_j$, and $\Gamma_j = (N_{\Gamma j}, z_{0j})$.

Drawbacks of R-TNCESs

The TNCES models for the four behavior modes of AAS are denoted by $\Gamma_1 = (N_{\Gamma 1}, z1_0)$, $\Gamma_2 = (N_{\Gamma 2}, z2_0)$, $\Gamma_3 = (N_{\Gamma 3}, z3_0)$, and $\Gamma_4 = (N_{\Gamma 4}, z4_0)$, respectively. The set of all possible reconfiguration events of AAS is marked by $\mathscr{R} = \{r_{1,s}, r_{2,s}, r_{3,s}, r_{4,s}, r_{1,s}^{-1}, r_{2,s}^{-1}, r_{3,s}^{-1}, r_{4,s}^{-1}, r_{1,2}, r_{1,3}, r_{1,4}, r_{2,4}, r_{3,4}\}$. The reconfiguration event $r_{i,}$ indicates a local reconfiguration that transforms robot Rbi into its energy-efficient mode and $r_{i,s}^{-1}$ is the reverse of $r_{i,s}$, that is, to transform robot Rbi from its energy-efficient mode into its working mode. The implementation of the events $r_{i,s}$ and $r_{i,s}^{-1}$ does not change the current behavior mode but can switch robot Rbi between its working mode and energy efficient mode according to its busy/idle status and waiting time. Finally, $r_{i,}$ $(i = j/\,)$ denotes a global reconfiguration event that transforms AAS from the configuration Mode i into Mode j.

The firing of a reconfiguration function of a R-TNCES changes the system configuration. As a consequence, if reconfiguration functions are applied to model local reconfiguration events for switching components between their working modes and energy-efficient modes directly, the number of system configurations should be enlarged. For example, configuration Mode 4 should be considered as four different configurations: (1) Both   and   are in their working modes, (2) is in working mode and is in energy-efficient mode, (3) is in working mode and is in energy-efficient mode, and (4) both   and   are in their energy-efficient modes. These four configurations are with the same structure. However, they should be verified separately. Obviously, this increases the verification cost and burdens the whole design process.

Generally, transitions in a R-TNCES model normal events of a reconfigurable discrete event control system, whereas reconfiguration functions are used to model system reconfiguration events. However, the concurrence of reconfiguration functions and transitions is not allowed in R-TNCESs, which is in fact inconsistent with requirements of REMSs. To make it clearer, let us take the modules , , and   as an example. Their TNCES-based models in Mode 1 and Mode 4 are shown in Figures 3 and 4, respectively. The differences between them are marked by dotted lines.

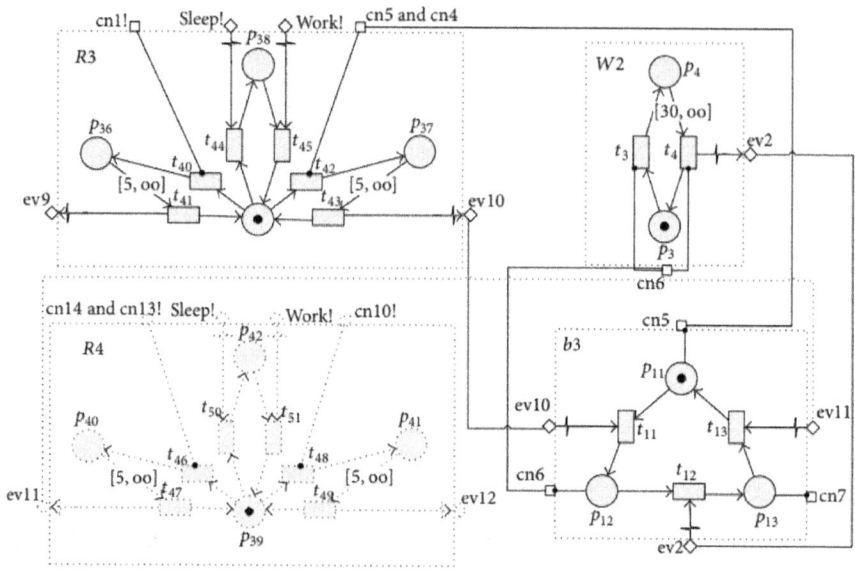

Figure 3: TNCES-based model of *R*3, *R*4, and *W*2 in Mode 1.

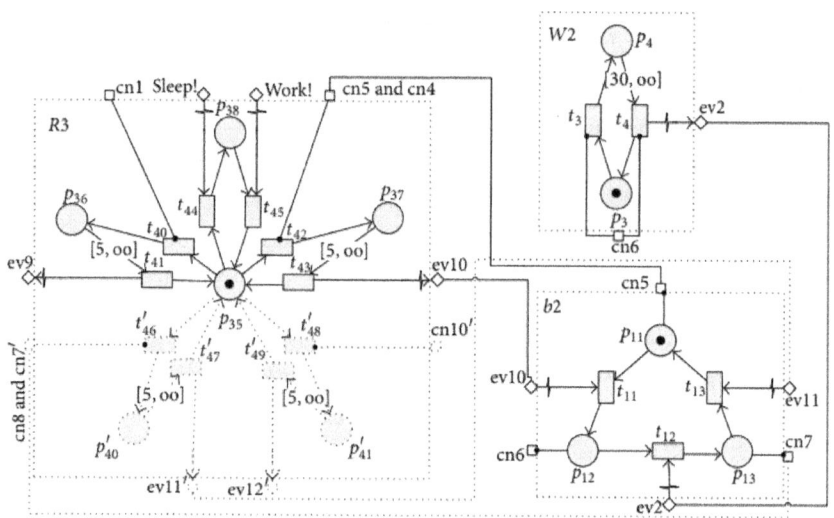

Figure 4: TNCES-based model of *R*3, *R*4, and *W*2 in Mode 4.

Example 3. Suppose that a reconfiguration function $r_{1,4}$ gets enabled at state *S*3 when AAS is in Mode 1. The physical meaning of *S*3 is that (1) Rb3 just finishes transporting S_A to $b2$ and (2) $W2$ is ready to process S_A. Assume that at this time a fault is detected in Rb4. Rb4 should be removed. Meanwhile, Rb3 must update itself soon in order to cover Rb4's task. According to the design requirements for AAS, $W2$ should go on working "naturally" at this time, that is, the enabled transition t_3 can fire at this state. However, the concurrence of reconfiguration functions and transitions is not allowed in R-TNCESs. Therefore, at state $S3$, only $r_{1,4}$ fires alone and AAS turns to the state $S1'$. Afterwards, t_3 fires, which leads to the next state $S2'$. However, if $r_{1,4}$ and t_3 fire together, AAS turns to the state $S2'$ directly without generating $S1'$.The state transition graph of this case is shown in Figure 5.

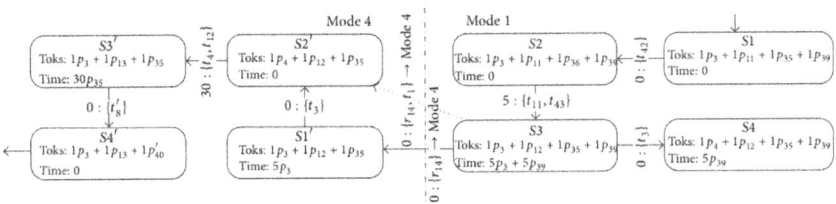

Figure 5: State transition graph of Example 3.

Example 4. Assume that two reconfiguration functions r_3 and $r_{4,s}$ get enabled simultaneously at state $S4$. The physical meaning of $S4$ is that (1) $W2$ just starts its work and (2) both Rb3 and Rb4 are idle. The firing of r_3 and $r_{4,s}$ only changes the states inside their modules but neither alters the system structure nor enables/disables any other transitions outside.That is to say, the firing of $r_{3,s}$ and $r_{4,s}$ does not change the current system configuration. According to the design requirements for AAS, both Rb3 and Rb4 can reconfigure themselves into energy-efficient modes freely when they are idle for more than two time units. However, the concurrence of multiple reconfiguration functions is not allowed in RTNCESs. Therefore, at state $S4$, only r_4 or $r_{3,s}$ fires alone. After that, the remaining one fires since it is still enabled. However, if r_3 and $r_{4,s}$ fire together, AAS turns to the state $S7'$ directly. The state transition graph of this case is shown in Figure 6.

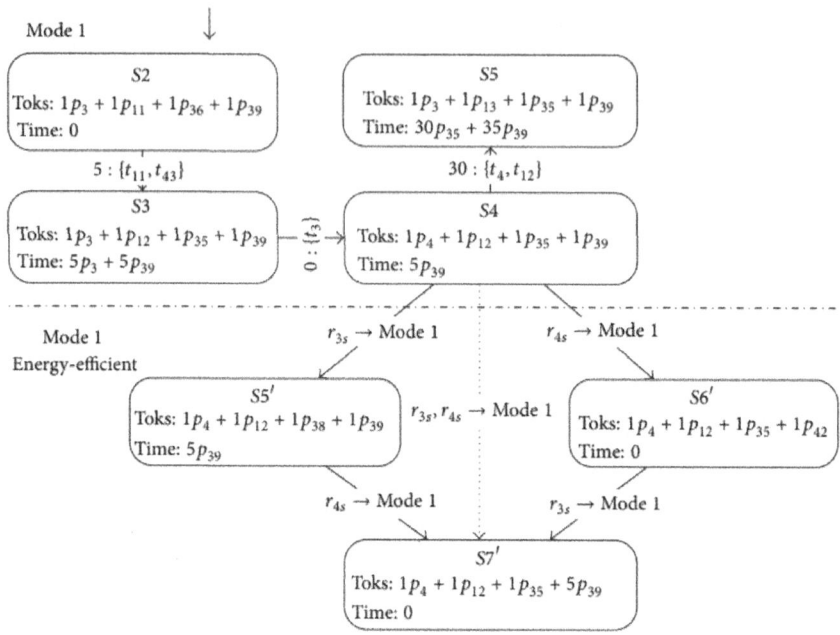

Figure 6: State transition graph of Example 4.

In conclusion, the original R-TNCESs are not sufficient to model a REMS. The reason can be explained from the following three aspects.(i) Reconfigurations at the component level only change component behavior modes between their working modes and energy-efficient modes rather than changing system configurations. If this kind of reconfigurations is modeled by reconfiguration functions directly, the number of system configurations should be enlarged, which increases the verification cost and burdens the whole design process.(ii)The concurrence of reconfiguration functions and transitions is not allowed in R-TNCESs. However, from the above examples, the concurrence of reconfiguration events and normal events is a common phenomenon in a REMS.(iii)Since the local reconfigurations for energy-efficient operations cannot be properly described, their corresponding dynamics and reasonable analysis cannot be performed.

To this end, this paper extends R-TNCESs to achieve two aims. First, all possible events including concurrent events that may occur in REMSs can be properly described. Second, the concurrence of reconfiguration functions and transitions should be controlled to ensure the system correctness.

Extended R-TNCESs

An extended R-TNCES has the same structure as the original R-TNCES. It is composed of a behavior module and a control module, denoted by eRN = {B, R}. The definition of system states is not changed, as shown in Definition 2 in Section 3.1. However, in the extended R-TNCES, reconfiguration functions are newly assigned with action ranges and concurrent decision functions. In addition, the firing rules of transitions and reconfiguration functions are updated such that they are conditionally allowed to fire concurrently.

Modified Reconfiguration Functions

In order to model the two types of reconfiguration events in a REMS directly, a concept, namely, action range, is developed for each reconfiguration function of a R-TNCES. In addition, a concurrent decision function is also assigned to a reconfiguration function to constrain concurrent transitions that may lead to undesired states such as deadlocks and overflow during a reconfiguration. For the sake of brevity, a reconfiguration function indicates a modified reconfiguration function in what follows.

Definition 5. A reconfiguration function r of an extended RTNCES eRN is a structure $r=$(Cond, s, x, Λ, Π). Cond $\rightarrow$ {true, false} is the precondition of $r.s$: $\Omega \rightarrow \Omega$ is the structure modification instruction. : $R(N_{\Gamma i}, z_{0i}) \rightarrow Z_{0j}$ is the state correlation function, where Z_{0j} is a set of feasible initial states of Γ_j. $\star r = \Gamma_i = (N_\Gamma, z_{0i})$ (resp., $r \star = \Gamma_j = (N_{\Gamma j}, z_{0j})$) denotes the TNCES before (resp., after) r fires. $\Lambda \in (N_{\Gamma i} \cup N_{\Gamma j})$ denotes the action range of r. $\Pi(r, Z) \rightarrow T$ is a concurrent decision function deciding a set of forbidden transitions that cannot fire together with r at state Z.

The reconfiguration functions of extended R-TNCESs are divided into two types: major and minor reconfiguration functions. For a reconfiguration function $r=$(Cond, s, x, Λ, Π) with $\star r = \Gamma_i = (N_{\Gamma i}, z_{0i})$ and $r \star = \Gamma_j = (N_{\Gamma j}, z_{0j})$, it is a major reconfiguration function if and only if $N_{\Gamma i} = N_{\Gamma j}$. Otherwise, it is a minor reconfiguration function. Let Rma and Rmi denote the sets of major and minor reconfiguration functions of eRN, respectively. Then we have R = $R_{ma} \cup R_{mi}$ and $R_{ma} \cap R_{mi} = 0$.

The implementation (firing) of a major reconfiguration function changes the structure of the current activated TNCES, whereas the implementation (firing) of a minor reconfiguration function only adjusts partial states of the activated TNCES within its action range.

Similar to Petri nets, the "conflict" concept is proposed for two enabled reconfiguration functions. We have the following two cases.

(1)   For two reconfiguration functions within the same type, that is, both being minor or major reconfiguration functions, if their action ranges have intersections, they are conflicting.

(2)   For a minor reconfiguration function and a major reconfiguration function, if the action range of the minor reconfiguration function is not completely covered by that of the major reconfiguration function, they are conflicting.

If two reconfiguration functions are conflicting, they cannot be implemented simultaneously. The symbol $r_1 \parallel r_2$ denotes that reconfiguration functions r_1 and r_2 are not conflicting. Similar to the definition of steps in TNCES, a r-step in an extended R-TNCES is a maximal set of reconfiguration functions that can fire simultaneously at a particular state. A r-step should satisfy the following two conditions.

(1)   For any two reconfiguration functions ri and r_j $(r_i \neq r_j)$ in a r-step γ, r_i and r_j are not conflicting; that is, $r_i \parallel r_j$.

(2)   There does not exist any other maximal set of reconfiguration functions γ' such that $\gamma \subset \gamma'$.

Accordingly, two r-steps γ_1 and γ_2 are conflicting, If $\exists r_1 \in \gamma_1$, $\exists r_2 \in \gamma_2, r_1 \neq r_2, r_1$ and r_2 are conflicting.

Dynamics of Extended R-TNCESs

Suppose that, at state $Z = [N_\Gamma, z]$, multiple reconfiguration functions get enabled, to be denoted by

$$R^* = \gamma_1 \cup \gamma_2 \cup \cdots \cup \gamma_g,$$

(4)

where γ_i $(i \in [1, g])$ is a maximal r-step at Z and, for all $i, j \in [1, g]$, $i = j$, γ_i and γ_j are conflicting. At the same state Z, the set of all enabled transitions is denoted by

$$T^* = u_1 \cup u_2 \cup \cdots \cup u_k,$$

(5)

where u_i $(i \in [1, k])$ is a maximal step and, for all $i, j \in [1, k]$, $i \neq j$, u_i and u_j are conflicting. For more information on how these steps are computed, please see [34, 35].

As a consequence, different compositions of r-steps and steps can occur simultaneously at this state. Given an enabled reconfiguration function r, we use $D \cdot T$ (resp., $D \cdot P$) to denote the set of deleted transitions (resp., deleted places) and $A \cdot T$ (resp., $A \cdot P$) to denote the set of added transitions (resp., added places) by firing it, where $^* r = \Gamma_i = (N_{\Gamma i}, z_{0i})$, $r^* = \Gamma_j = (N_{\Gamma j}, z_{0j})$, $N_{\Gamma i} = (P_i, T_i, F_i, CN_i, EN_i, em_i, DC_i)$, and $N_{rj} = (P_j, T_j, F_j, CN_j, EN_j, em_j, DC_j)$. We have the following two cases.

(1)   For a transition $t \in T_i$, if it is enabled simultaneously with a minor reconfiguration function r=(Cond, s, x, Λ, Π) at state $\mathcal{Z} = \lceil N_r, z \rfloor$ and $t \notin \Lambda_r$, then t can fire simultaneously with r; that is, $t \notin \Pi(r, Z)$.

(2)   For a transition $t \in T_i$, if it is enabled simultaneously with a major reconfiguration function r=(Cond, s, x, Λ, Π) at state $Z = \lceil N_r, z \rfloor$, then we have the following two subcases.

(A) A spontaneous transition t is forbidden to be concurrent with r at Z, if it meets one of the following conditions.

(i)   If it is deleted by r, that is, $t \in D \cdot T$, it is forbidden by r; that is, $t \in \Pi(r, Z)$.

(ii)   If $t \notin D \cdot T$ and all its elements are not changed by firing r, then it is allowed to fire simultaneously with r. Formally, if ${}^\bullet t_i = {}^\bullet t_j, t_i^\bullet = t_j^\bullet, {}^- t_i = {}^- t_j, {}^\sim t_i = {}^\sim t_j = \emptyset$, and $em(t)_i = em(t)_j$, we have $t \notin \Pi(r, \mathcal{Z})$.

(iii) If $t \notin D \cdot T$, some of its elements are modified by r, which include its preset, postset, source places, and firing mode, and we have the following two cases.

(a)   The preset, source places, and firing mode of t decide whether t is enabled after the firing of r. Therefore, if its preset, source places, or firing mode is changed by r, it can fire simultaneously with r. Formally, if ${}^\bullet t_i \neq {}^\bullet t_j, {}^- t_i \neq {}^- t_j,$ or $em(t)_i \neq em(t)_j$, then $t \notin \Pi(r, \mathcal{Z})$.

(b)   The postset of t does not change its enabling condition but influences the structure of the net. Therefore, it is forbidden by r. Formally, if $t_i^\bullet \neq t_j^\bullet$, we have $t \in \Pi(r, \mathcal{Z})$.

(B) A forced transition t is forbidden to be concurrent with r at Z, if it further meets one of the following conditions.

(i)   Its firing mode is - and all of its forcing transitions are forbidden to be concurrent with r; that is, if $em(t) = em_j(t) = -$ and, for all $t' \in {}^\sim t, t' \notin \Pi(r, \mathcal{Z})$, then $t \in \Pi(r, \mathcal{Z})$.

(ii)   Its firing mode is , and at least one of its forcing transitions is forbidden by r; that is, if $em_i(t) = em_j(t) = $, and $t \in \Pi(r, \mathcal{Z})$.

Since an extended R-TNCES allows the concurrence of multiple reconfiguration functions and transitions, the reachability graph of an extended R-TNCES is defined as follows.

Definition 6. The reachability graph of an extended R-TNCES is a combination of several labeled directed graphs whose nodes are the states of and whose arcs are of three kinds: steps, -steps, and combinations of a step and a -step.

(i)   The arc from state $\lceil N_{\Gamma i}, z_i \rfloor$ to state $\lceil N_{\Gamma i}, z'_i \rfloor$ is $z'_i \rfloor$, where $z'_i \in R(N_{\Gamma i}, z_i)$

(ii)   The arc from state $[N_\Gamma, z_i]$ to state $[N_{\Gamma j}, z_j]$ is labeled with a r-step γ represented by $\lceil N_{\Gamma i}, z_i][\gamma\rangle[N_{\Gamma j}, z_j]$, if γ contains major reconfiguration functions with $N_{\Gamma i} = N \Gamma j$. Otherwise, we have $N_{\Gamma i} = N_{\Gamma j}$ and $z_j \notin (N_{\Gamma i}, z_i)$.

(iii) The arc from $[N_\Gamma, z_i]$ to state $[N_{\Gamma j}, z_j]$ is labeled with a step and a r-step $\{R, u\}$ represented by $[N_{\Gamma i}, z_i][\gamma, u\rangle[N_{\Gamma j}, z_j]$, if γ contains major reconfiguration functions with $N_{\Gamma i} = N_{\Gamma j}$. Otherwise, we have $N_{\Gamma i} = N_{\Gamma j}$.

Obviously, the graphical representation of an extended R-TNCES model is the same as that of a R-TNCES model. However, system dynamics get enriched along with the changes of reconfiguration functions. If we use an extended R-TNCES to model AAS, the graphical TNCES models shown in Figures 3 and 4 are still correct. However, their reachability graphs get enriched during same reconfiguration.

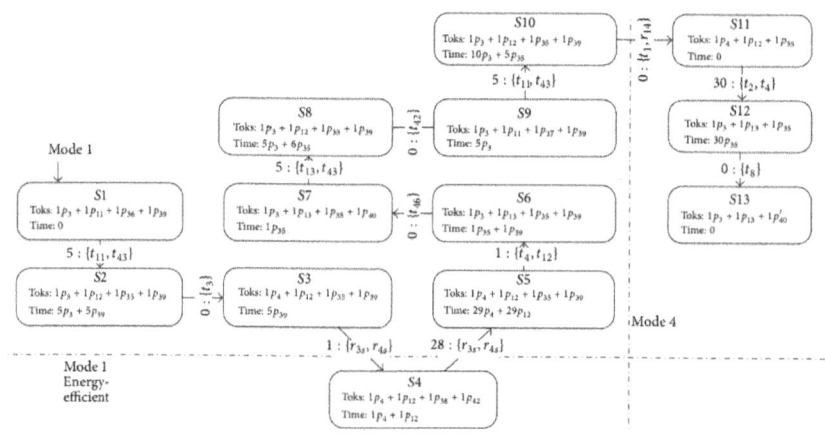

Figure 7: A fragment of the reachability graph of Example 7.

Example 7. A fragment of the reachability graph of the extended R-TNCES-based model of the example composed of Rb3, Rb4, and $W2$ is shown in Figure 7. AAS starts running in *Mode* 1. When it arrives at state $S3$, two minor reconfiguration functions $r_{3,}$ and $r_{4,s}$ get enabled and fire simultaneously to reconfigure robots Rb3 and Rb4 into their energy-efficient modes. After 28 time units, they reconfigure back to working modes. Assume that $R4$ is detected to have a fault at state $S10$, the major reconfiguration function $r_{1,4}$ gets enabled.

In the meantime,$_3$ gets enabled simultaneously with $r_{1,4}$. Therefore, t_3 fires simultaneously with $r_{1,4}$, which leads to the transformation of AAS into *Mode* 4.

VERIFICATION OF REMSS BASED ON EXTENDED R-TNCESS

In order to perform correct formal verification of AAS, an extended R-TNCES-based model should be built for it. The extended R-TNCES based model of AAS is marked by $^e\text{RN}_{\text{AAS}} = \{\mathcal{B}, \mathcal{R}\}$, $\mathcal{B} = \Gamma_1 \cup \Gamma_2 \cup \Gamma_3 \cup \Gamma_4$, $\mathcal{R} = \mathcal{R}_{ma} \cup \mathcal{R}_{mi}$, $\mathcal{R}_{ma} = \{r_{1,2}, r_{1,3}, r_{1,4}, r_{2,4}, r_{3,4}\}$, and $\mathcal{R}_{mi} = \{r_{1,s}, r_{2,s}, r_{3,s}, r_{4,s}, r_{1,s}^{-1}, r_{2,s}^{-1}, r_{3,s}^{-1}, r_{4,s}^{-1}\}$. We have $\Omega = \{N_{\Gamma_1}, N_{\Gamma_2}, N_{\Gamma_3}, N_{\Gamma_4}\}$. The four major reconfiguration functions are conflicting with each other. The minor reconfiguration n functions r_i and $r_{i,s}^{-1}$ ($i \in [1,4]$) are conflicting but others are not. The behavior module of $^e\text{RN}_{\text{AAS}}$ is shown in Figure 8, where elements drawn by dotted lines are possibly modified during the implementation of a major reconfiguration function. In order to apply automatic model checking to an extended R-TNCES, a TNCES-based nested state machine is developed to implement its control module.

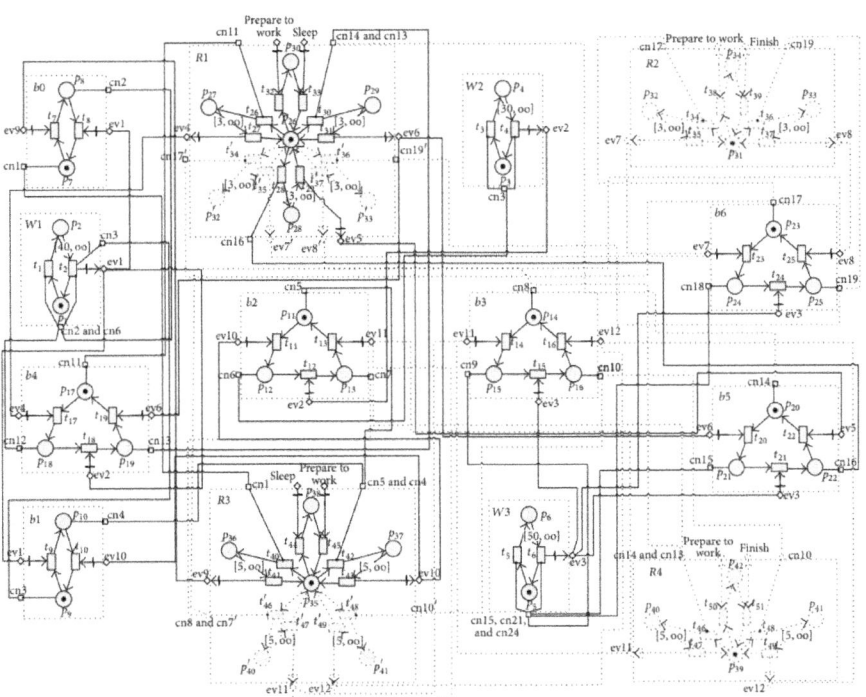

Figure 8: Behavior module of eRN$_{\text{AAS}}$.

Implementation of Extended R-TNCESs

First of all, major reconfiguration functions are grouped according to their action ranges. A set of state machines specified by TNCESs, which are called s, is defined. Each state machine corresponds to a group of major reconfiguration functions that share the same action range. In a particular , each transition corresponds to a major reconfiguration function. The transitions in a state machine cannot fire simultaneously, which means that these modeled major reconfiguration functions by one state machine are conflicting with each other. Firing a transition in a implies that a major reconfiguration function is implemented. A Major -Charger is formalized as follows:

$$\text{Major\_changer} = (P, T, F, em, z_0),$$

(6)

where, for any $t \in T, |{}^\bullet t| = |t^\bullet| = 1, \sum M_0(P) = 1$, which means that only one place in P owns a token at the initial state, and $em : T \to \{ \boxtimes \}$. The precondition Cond can be modeled by input event/condition signals from external to transitions in a Major changer.

In addition, an actuator denoted by Actuator is defined for each place sp in all Major changers, which is marked by Actuator = Act(sp). Each actuator is composed of a place ap and a transition at only, where $\cdot \, {}^\bullet ap = ap^\bullet = \{at\}, \, {}^\bullet at = at^\bullet = = \{ap\}$, and $(ap) = 1$. When the place sp in a Major changer receives a token, the actuator Actuator = Act(sp) is activated. An Actuator is formalized as follows:

$$\text{Actuator} = (P, T, F, em, z_0),$$

(7)

where $|P| = |T| = 1, \, {}^\bullet at = at^\bullet = \{ap\}, \, {}^\bullet ap = ap^\bullet = \{at\}, \, m_0(P) = 1$, and $em : T \to \{ \boxtimes \}.$.

Similar to major reconfiguration functions, minor reconfiguration functions are grouped according to their action ranges. A set of state machines specified by TNCESs, which are called Minor changers, is defined. Each state machine corresponds to a group of minor reconfiguration functions. If the action ranges of two minor reconfiguration functions are the same, they are modeled by transitions in a Minor changer. If the action range of a group of minor reconfiguration functions, to be modelled by a Minor changer, is completely covered by that of a group of major reconfiguration functions, to be modeled by a Major changer, then this Minor changer is activated while this Major changer is activated.

A Minor_changer is formalized as follows:

$$\text{Minor\_changer} = (P, T, F, em, z_0),$$

(8)

where, for any $t \in T, |{}^\bullet t| = |t^\bullet| = 1, \sum M_0(P) = 1$, which means that only one place in P owns a token at the initial state, and $em \; T \to \{ \boxtimes \}.$ The precondition

Cond can be modeled by input event/condition signals from external to transitions in a Minor changer.

Example 8. Figure 9 depicts the TNCES-based control module of eRN_{AAS}. It has only one Major changer, since the four major reconfiguration functions share the same action range. It has four Minor changers, since the four robots have four distinguished action ranges. Places p_1, p_2, p_3, and p_4 in Major changer correspond to Mode 1, Mode 2, Mode 3, and Mode 4, respectively. When t_3 fires, the major reconfiguration function $r_{1,4}$ is implemented. Robots Rb3 and Rb1 are applied in every mode of AAS. Therefore, minor reconfiguration functions that transform them between energyefficient modes and working modes are activated in every system behavior mode. Moreover, it is possible for them to fire simultaneously with other major reconfiguration functions.

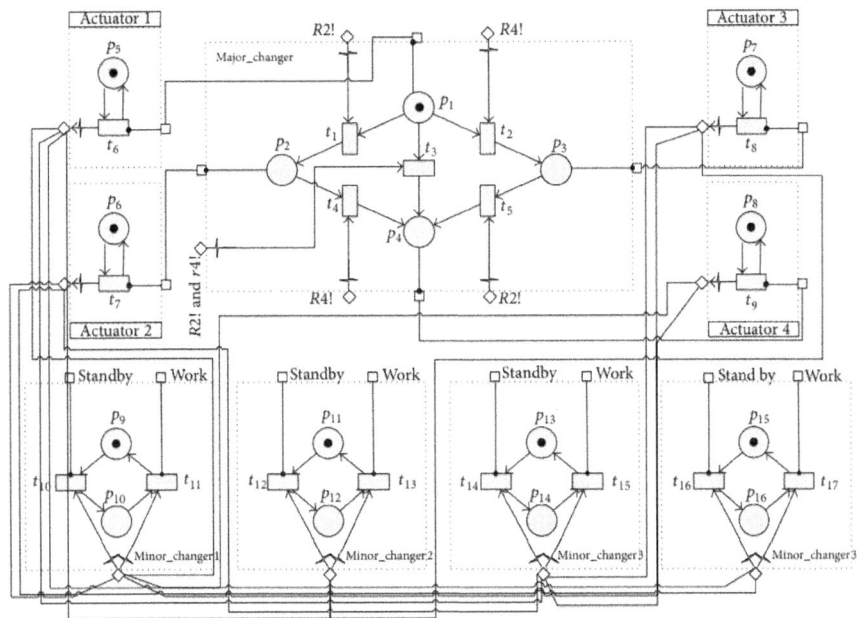

Figure 9: Control module of eRN_{AAS}.

Formal Verification of AAS

Since the time when a major reconfiguration function can get enabled and fire cannot be predicted, this paper applies an instruction insertion method to simulate AAS. In addition, eRNAAS evolves according to fired maximal steps and r-steps. Assume that AAS should finish 100 subassemblies. It starts with Mode 1. At time t_1 when it finishes the 60th subassembly, it reconfigures into

Mode 2 due to the fault detection of Rb2. Then, it goes on working in Mode 2. At time t_2 when the 91st subassembly is being processed, it transforms into Mode 4 according to the fault detection of Rb4. During the whole process, minor reconfigurations, that is, transforming robots between their working modes and energy-efficient modes, are applied.

SESA is applied to compute the reachability graph of this whole process. A minimal path regarding time consumption from the initial state to the objective state is computed in each mode. In Mode 1, it generates 23044 states, taking 6990 time units to finish assembly of the first 60 subassemblies in the minimal path. In Mode 2, it generates 85259 states, costing 4127 time units to finish assembling the next 30 subassemblies in the minimal path. Finally, in Mode 4, it generates 195007 states, taking 1525 time units to finish assembling the last 10 subassemblies in the minimal path. Note that two states can be considered to be same if and only if they have the same token numbers and time status.

Since each TNCES-based model of the behavior modes of AAS is a well-designed control system, they are proved to be qualified according to SESA, where eCTL based functional properties and TCTL based temporal properties are checked. In addition, the following eCTL formula is applied to the control module of eRNAAS:

$$Z_0 = EX \langle t4ANDt12 \rangle X \langle p12 = 1 \rangle . \tag{9}$$

This formula is proved to be false by SESA. Transition t_{12} corresponds to minor reconfiguration function r_2. Therefore, it can fire only when AAS is in Mode 1 or Mode 2.The following formula is proved to be true:

$$Z_0 = EX \langle t2ANDt10 \rangle X \langle p10 = 1 \rangle . \tag{10}$$

It means that when robot Rb4 breaks down, two reconfiguration functions $r_{1,3}$ and $r_{1,s}$ are possible to fire simultaneously.

The triggering conditions of minor reconfiguration functions can be computed previously. There are several possible state/event paths showing system behavior from the initial state to the objective state, at which 100 subassemblies are finished. We select a minimal path regarding time for each TNCES-based model of the three configurations, to be denoted by Path = $Z_1, Z_2,..., Z_n$, where energy-efficient operations are not included. That is to say, all robots should stay in their working modes in this case although they should wait for a period of time before the next task comes. After that, based on the states on this path, the time when a minor reconfiguration function gets enabled and fires can be computed. For example, if an activated robot starts to wait at a particular state Z_i, at which the system time is τ_1, a search is performed along this minimal path at τ_1. If it is found that at Z_j the robot works again, at which the system time is τ_2, then the time delay $\Delta_\tau = \tau_2 - \tau_1$ between these two states is obtained. The round local reconfigurations for switching a robot between its working mode and energy-efficient mode take two time units. Therefore, if the time delay is larger than two, that is, $\Delta_\tau > 2$, a local reconfiguration can be applied to this robot. The system time for reconfiguring this robot from its working mode to its energy-efficient mode is τ_1. The system time for reconfiguring this robot from its energyefficient mode to its working mode is $\tau_2 - 1$.

The time of robots on their energy-efficient modes in minimal paths is computed during the assembly of 100 subassemblies. They are shown in Table 1 together with the whole system uptime in each mode. Take Mode 1 as an example. Assume that Rb1 consumes one energy unit per time unit in its working time but only consumes 30% energy unit per time unit in its energy-efficient mode. In *Mode* 1, if there is no minor reconfiguration applied to Rb1 for saving energy, it will consume 6990 energy units. However, it only consumes $6990 - 3233 + 30\% \times 3233 = 4726.9$ energy units in Mode 1 if minor reconfigurations are applied when it is idle. In the same way, the energy saved by the robots during this simulation is shown in Table 2, where the third row shows the energy consumption of each robot if no minor reconfigurations are applied, the fourth row shows the energy consumption of each robot when minor reconfigurations are applied, and the last row shows the saved energy of each robot during this process.

Table 1: Time of robots on their energy-efficient modes

Configuration			Mode 1			Mode 2		Mode 4	
System uptime			6690			4127		1525	
Robot	Rb1	Rb2	Rb3	Rb4	Rb1	Rb3	Rb4	Rb1	Rb3
Time on energy-efficient mode	3233	4455	3818	2643	1004	2458	2428	523	435

Table 2: Energy consumption of robots

Configuration			Mode 1			Mode 2		Mode 4	
Robot	Rb1	Rb2	Rb3	Rb4	Rb1	Rb3	Rb4	Rb1	Rb3
Energy 1	6690	6690	6690	6690	4127	4127	4127	1525	1525
Energy 2	4726.9	3871.5	4303.4	5139.9	3424.2	2406.4	2427.4	1158.9	1220.5
Saved energy	1963.1	2818.5	2386.6	1550.1	702.8	1720.6	1699.6	366.1	304.5

CONCLUSION

A reconfigurable and energy-efficient manufacturing system (REMS) is a typical reconfigurable discrete event control system. It allows two kinds of dynamic system reconfigurations: local and global reconfigurations. The former ones are applied to save energy for components, whereas the latter ones are applied to change system configurations according to changed inner/outer execution environments. Meanwhile, normal events should be conditionally allowed to occur simultaneously with these system reconfigurations, such that the system can reconfigure smoothly and safely. In order to easily model conditioned concurrence of reconfiguration events and normal events and represent all interesting system behavior, this paper extends the reconfigurable timed net condition event systems (R-TNCESs) formalism. Original reconfiguration functions are newly assigned with action ranges and concurrent decision functions. Accordingly, the dynamics of R-TNCES is updated. After that, a TNCES-based implementation method for the proposed extended R-TNCES is developed such that automatic model checking can be applied. The verified properties include functional, temporal, and energy properties that are specified by Computation Tree Logic (CTL), extended Computation Tree Logic (eCTL), or Timed Computation Tree Logic (TCTL). An automatic assembly system is used to illustrate the whole work.

In the future, the authors will focus on reasonably optimal reconfigurable control systems that can save more energy and the applications of the proposed method to the crude-oil operation enterprises with huge energy consumption [48].

ACKNOWLEDGMENTS

This work was supported in part by the National Natural Science Foundation of China under Grant no. 61374068 and the Science and Technology Development Fund, MSAR, under Grant nos. 065/2013/A2 and 066/2013/A2.

REFERENCES

1. Y. Koren, U. Heisel, F. Jovane et al., "Reconfigurable manufacturing systems," CIRP Annals—Manufacturing Technology, vol. 48, no. 2, pp. 527–540, 1999. ·

2. M. Khalgui, O. Mosbahi, J. F. Zhang, Z. W. Li, and A. Gharbi, "Feasible dynamic reconfigurations of petri nets: application to a production systems," in Proceedings of the 6th International Conference on Software and Database Technologies (ICSOFT ‹11), pp. 105–110, Sevilla, Spain, July 2011.

3. T. Parisini and S. Sacone, "Fault diagnosis and controller re-configuration: an hybrid approach," inProceedings of the IEEE International Symposium on Intelligent Control (ISIC ‹98), pp. 163–168, September 1998.

4. P. Leitão, J. Alves, J. M. Mendes, and A. W. Colombo, "Energy aware knowledge extraction from petri nets supporting decision-making in

service-oriented automation," in Proceedings of the IEEE International Symposium on Industrial Electronics (ISIE ‹10), pp. 3521–3526, Bari, Italy, July 2010.

5. S. Karnouskos, A. W. Colombo, J. L. M. Lastra, and C. Popescu, "Towards the energy efficient future factory," in Proceedings of the 7th IEEE International Conference on Industrial Informatics (INDIN ‹09), pp. 367–371, Cardiff, Wales, June 2009.

6. K. Bunse, M. Vodicka, P. Schönsleben, M. Brülhart, and F. O. Ernst, "Integrating energy efficiency performance in production management— gap analysis between industrial needs and scientific literature," Journal of Cleaner Production, vol. 19, no. 6-7, pp. 667–679, 2011.

7. S. Mechs, S. Lamparter, and J. P. Müller, "On evaluation of alternative switching strategies for energy-efficient operation of modular factory automation systems," in Proceedings of the IEEE 17th International Conference on Emerging Technologies and Factory Automation (ETFA ‹12), pp. 1–8, IEEE, Kraków, Poland, September 2012.

8. S. Mechs, J. P. Muller, S. Lamparter, and J. Peschke, "Networked priced timed automata for energy-efficient factory automation," in Proceedings of the American Control Conference (ACC ‹12), pp. 5310–5317, Montreal, Canada, June 2012.

9. Z. M. Bi and L. Wang, "Optimization of machining processes from the perspective of energy consumption: a case study," Journal of Manufacturing Systems, vol. 31, no. 4, pp. 420–428, 2012.

10. Y. Oda, Y. Kawamura, and M. Fujishima, "Energy consumption reduction by machining process improvement," Procedia CIRP, vol. 4, pp. 120–124, 2012.

11. Cannata, S. Karnouskos, and M. Taisch, "Energy efficiency driven process analysis and optimization in discrete manufacturing," in Proceedings of the 35th Annual Conference of the IEEE Industrial Electronics Society(IECON ‹09), pp. 4449–4454, Porto, Portugal, November 2009.

12. G. Mouzon and M. B. Yildirim, "A framework to minimise total energy consumption and total tardiness on a single machine," International Journal of Sustainable Engineering, vol. 1, no. 2, pp. 105–116, 2008. ·

13. D. Shorin and A. Zimmermann, "Model-based development of energy-efficient automation systems," inProceedings of the 17th IEEE Real-Time and Embedded Technology and Applications Symposium (RTAS ‹11), Chicago, Ill, USA, April 2011.

14. C.-W. Park, K.-S. Kwon, W.-B. Kim et al., "Energy consumption reduction technology in manufacturing—a selective review of policies,

standards, and research," International Journal of Precision Engineering and Manufacturing, vol. 10, no. 5, pp. 151–173, 2009.

15. P. Stoffels, W. M. Boussahel, M. Vielhaber, and G. Frey, "Energy engineering in the virtual factory," inProceedings of the IEEE 18th International Conference on Emerging Technologies and Factory Automation (ETFA ‹13), pp. 1–6, Cagliari, Italy, September 2013.

16. N. Wu, M. Zhou, and Z. Li, "Resource-oriented Petri net for deadlock avoidance in flexible assembly systems," IEEE Transactions on Systems, Man, and Cybernetics Part A:Systems and Humans, vol. 38, no. 1, pp. 56–69, 2008.

17. Z. Li and M. Zhou, "Two-stage method for synthesizing liveness-enforcing supervisors for flexible manufacturing systems using Petri nets," IEEE Transactions on Industrial Informatics, vol. 2, no. 4, pp. 313–325, 2006.

18. Z. W. Li, H. S. Hu, and A. R. Wang, "Design of liveness-enforcing supervisors for flexible manufacturing systems using Petri nets," IEEE Transactions on Systems, Man and Cybernetics Part C: Applications and Reviews, vol. 37, no. 4, pp. 517–526, 2007.

19. Z. Li and M. Zhou, "Control of elementary and dependent siphons in Petri nets and their application,"IEEE Transactions on Systems, Man, and Cybernetics Part A: Systems and Humans, vol. 38, no. 1, pp. 133–148, 2008.

20. Z. Li, M. Zhou, and N. Wu, "A survey and comparison of Petri net-based deadlock prevention policies for flexible manufacturing systems," IEEE Transactions on Systems, Man and Cybernetics Part C: Applications and Reviews, vol. 38, no. 2, pp. 173–188, 2008. ·

21. Z. Li, N. Wu, and M. Zhou, "Deadlock control of automated manufacturing systems based on petri nets—a literature review," IEEE Transactions on Systems, Man and Cybernetics Part C: Applications and Reviews, vol. 42, no. 4, pp. 437–462, 2012.

22. Z. Y. Ma, Z. W. Li, and A. Giua, "Design of optimal Petri net controllers for disjunctive generalized mutual exclusion constraints," IEEE Transactions on Automatic Control, 2015.

23. J. H. Ye, Z. W. Li, and A. Giua, "Decentralized supervision of Petri nets with a coordinator," IEEE Transactions on Systems, Man, and Cybernetics: Systems, vol. 45, no. 6, pp. 955–966, 2015.

24. Z. W. Li, M. C. Zhou, and M. D. Jeng, "A maximally permissive deadlock prevention policy for FMS based on petri net siphon control and the theory of regions," IEEE Transactions on Automation Science

and Engineering, vol. 5, no. 1, pp. 182–188, 2008.

25. Z. W. Li, G. Y. Liu, H.-M. Hanisch, and M. C. Zhou, "Deadlock prevention based on structure reuse of petri net supervisors for flexible manufacturing systems," IEEE Transactions on Systems, Man and Cybernetics, Part A: Systems and Humans, vol. 42, no. 1, pp. 178–191, 2012.

26. Y. F. Chen and Z . W. Li, "Design of a maximally permissive liveness-enforcing supervisor with a compressed supervisory structure for flexible manufacturing systems," Automatica, vol. 47, no. 5, pp. 1028–1034, 2011.View at MathSciNet ·

27. Y. F. Chen, Z. W. Li, M. Khalgui, and O. Mosbahi, "Design of a maximally permissive liveness- enforcing Petri net supervisor for flexible manufacturing systems," IEEE Transactions on Automation Science and Engineering, vol. 8, no. 2, pp. 374–393, 2011.

28. Y. F. Chen, Z. W. Li, K. Barkaoui, and M. Uzam, "New Petri net structure and its application to optimal supervisory control: Interval inhibitor arcs," IEEE Transactions on Systems, Man, and Cybernetics: Systems, vol. 44, no. 10, pp. 1384–1400, 2014.

29. X. Wang, I. Khemaissia, M. Khalgui, Z. Li, O. Mosbahi, and M. Zhou, "Dynamic low-power reconfiguration of real-time systems with periodic and probabilistic tasks," IEEE Transactions on Automation Science and Engineering, vol. 12, no. 1, pp. 258–271, 2015.

30. J. Zhang, M. Khalgui, Z. Li, O. Mosbahi, and A. M. Al-Ahmari, "R-TNCES: a novel formalism for reconfigurable discrete event control systems," IEEE Transactions on Systems, Man, and Cybernetics: Systems and Humans, vol. 43, no. 4, pp. 757–772, 2013.

31. J. F. Zhang, M. Khalgui, Z. W. Li, G. Frey, O. Mosbahi, and H. B. Salah, "Reconfigurable coordination of distributed discrete event control systems," IEEE Transactions on Control Systems Technology, vol. 23, no. 1, pp. 323–330, 2015.

32. C. Gerber, S. Preuße, and H.-M. Hanisch, "A complete framework for controller verification in manufacturing," in Proceedings of the 15th IEEE International Conference on Emerging Technologies and Factory Automation (ETFA ‹10), pp. 1–9, September 2010. ·

33. C. Gerber, Implementation and Verification of Distributed Control Systems, Logos, Berlin, Germany, 2011.

34. H.-M. Hanisch, J. Thieme, A. Lueder, and O. Wienhold, "Modeling of PLC behavior by means of timed net condition/event systems," in Proceedings of the IEEE 6th International Conference on Emerging Technologies and Factory Automation (ETFA ‹97), pp. 391–396, IEEE,

Los Angeles, Calif, USA, September 1997.

35. M. Rausch and H. M. Hanisch, "Net condition/event systems with multiple condition outputs," inProceedings of the 1995 INRIA/IEEE Symposium on Emerging Technologies and Factory Automation, pp. 592–600, Paris, France, October 1995.

36. T. Murata, "Petri nets: properties, analysis and applications," Proceedings of the IEEE, vol. 77, no. 4, pp. 541–580, 1989.

37. Y. F. Chen, Z. W. Li, and M. C. Zhou, "Optimal supervisory control of flexible manufacturing systems by Petri nets: a set classification approach," IEEE Transactions on Automation Science and Engineering, vol. 11, no. 2, pp. 549–563, 2014.

38. Z. W. Li and M. C. Zhou, "Elementary siphons of Petri nets and their application to deadlock prevention in flexible manufacturing systems"," IEEE Transactions on Systems, Man, and Cybernetics Part A: Systems and Humans., vol. 34, no. 1, pp. 38–51, 2004.

39. Z. W. Li and M. C. Zhou, "Clarifications on the definitions of elementary siphons in Petri nets," IEEE Transactions on Systems, Man, and Cybernetics, Part A: Systems and Humans, vol. 36, no. 6, pp. 1227–1229, 2006.

40. Z. Li and M. Zhao, "On controllability of dependent siphons for deadlock prevention in generalized Petri nets," IEEE Transactions on Systems, Man, and Cybernetics Part A: Systems and Humans, vol. 38, no. 2, pp. 369–384, 2008.

41. P. H. Starke and S. Roch, "Analysing signal-net systems," Tech. Rep., Informatik Berichte, Humboldt-University, Berlin, Germany, 2002.

42. E. M. Clarke, O. Grumberg, and D. Peled, Model Checking, MIT Press, 1999.

43. E. A. Emerson, A. K. Mok, A. P. Sistla, and J. Srinivasan, "Quantitative temporal reasoning," in Computer-Aided Verification, vol. 531 of Lecture Notes in Computer Science, pp. 136–145, Springer, Berlin, Germany, 1991.

44. S. Roch, "Extended computation tree logic," in Proceedings of the Informatik-Bericht Workshop on Concurrency, Specification and Programming, pp. 225–234, 2000.

45. S. Preuße, D. Missal, C. Gerber, M. Hirsch, and H. M. Hanisch, "On the use of model-based IEC 61499 controller design," International Journal of Discrete Event Control Systems, vol. 1, no. 1, pp. 115–128, 2011.

46. S. Preuse, H.-C. Lapp, and H.-M. Hanisch, "Closed-loop system

modeling, validation, and verification," in Proceedings of the IEEE 17th International Conference on Emerging Technologies and Factory Automation (ETFA ‹12), pp. 1–8, IEEE, Kraków, Poland, September 2012.

47. S. Preußse and H.-M. Hanisch, "Verifying functional and non-functional properties of manufacturing control systems," in Proceedings of the 3rd International Workshop on Dependable Control of Discrete Systems (DCDS ‹11), pp. 41–46, Saarbrücken, Germany, June 2011.

48. N. Q. Wu, M. C. Zhou, and Z. W. Li, "Short-term scheduling of crude-oil operations: petri net-based control-theoretic approach," IEEE Robotics & Automation Magazine, 2015.

Chapter 3

ANALYSIS, MODELING AND SIMULATION OF A POLY-BAG MANUFACTURING SYSTEM

R. A. R. C. Gopura[1], T. S. S. Jayawardane[2]

[1]Department Mechanical Engineering, University of Moratuwa, Kaubedda, Sri Lanka

[2]Department of Textile and Clothing Technology, University of Moratuwa, Kaubedda, Sri Lanka

ABSTRACT

The cost of raw material of poly-bags increases and fluctuates with an unpredictable trend. Further, legal restrictions imposed on some types of polythene products adversely affects for the demand. In this context, entrepreneurs engaging in poly-bag manufacturing face major challenges. With the purview of optimizing the poly-bag manufacturing process, authors attempted to analyze, model and simulate the poly-bag manufacturing process in the light of posed challenges. This paper presents preliminary analysis, modeling and simulation strategies of a poly-bag manufacturing system. In addition, a risk prioritization method is proposed in the preliminary analysis and also a simulation tool is developed.

INTRODUCTION

The raw material prices of poly-bags increase continuously with a high degree of fluctuation. Hence, the forecasting of material prices with adequate accuracy is a really challenge. Since the level of price fluctuation is so severe, sometimes it is lucrative to purchase raw materials in massive quantities during the low price periods and store them for the future use. The backward integration is quite difficult to achieve with the amount of raw material purchases, and difficulty in predicting the raw material prices with adequate accuracy make the situation worse. In addition, legal restrictions imposed on some types of polythene products have a serious adverse effect on the demand for the products.

Therefore, any constructive contribution in the poly-bag manufacturing value chain has a great impact to its productivity. When switching from one product to another, parameters in poly-bag manufacturing system have to be varied and it is associated with a considerable set up time as well as a substantial amount of raw material waste. The fluctuation of raw material prices has caused serious problems in current inventory control practice and it is a great barrier to adapt popular lean manufacturing techniques in the production system. In order to face the posed challenges successfully, a poly-bag manufacturing system needs to have an in depth analysis of the production process, so that solutions can be recommend in the light of posed challenges.

Various categories of polyethylene [i.e., high density polyethylene (HDPE), low density polyethylene (LDPE)and linear low density polyethylene (LLDPE)] and polypropylene (PP) are basically used as raw materials for the poly-bag manufacturing [1]. The main process is the film blowing [1,2]. It was followed by subsequent processes namely cutting, sealing, printing, quality checking (QC), and packing. The majority of the research in the area of poly-bag manufacturing has been concerned in a single process: the film blowing [3-5]. In the film blowing, a significant work has been carried out to show interaction of the polymer rheology with the process [4]. Relatively, little work has been done in modeling and simulation of poly-bag manufacturing system. However, few researchers have been tried out to simulate manufacturing processes [6-8]. Brown and his colleagues [6] implemented performance modeling capability (simulation, capacity analysis, and cost analysis) at factories of Siemens Semiconductor for both wafer fabrication and back-end operations. Graul et al., 2003 [7] presented a concept and a framework to capture and maintain the multiple descriptions and its applicability in modeling and simulation of manufacturing systems. They explained a knowledgebased approach to support the integration of multiple descriptions with collected data from legacy status for the use in the design and generation of valid simulation models. Starting with a short analysis of the current situation in the field of factory simulation and an overview of current tendencies in the manufacturing area, Schumann [8] introduced a method to integrate High Level Architecture (HLA) and existing simulation tools. They presented the simulation tool SLX [9] and the visualization tool Skopeo [10], which were both utilized to perform a prototype federation of a manufacturing plant.

In this study, initially a typical poly-bag manufacturing system is analyzed [11] to identify strengths, weaknesses, opportunities, and threats of the system. Further, risks and bottlenecks analysis were carried out in the preliminary stage to identify the risks and bottlenecks of the selected poly-bag manufacturing system. A risk prioritization method is also proposed in the risk analysis. The risks are prioritized based on their effects for the system. Then a mathematical model of the poly-bag manufacturing system is developed with reasonable assumptions to obtain the optimum throughput time of a given type of a poly-bag under available plant capabilities and to minimize the material wastage under current production setup. The system is simulated using the mathematical model developed. The production schedule for maximum productivity is generated for the input customer orders. For simulation, a graphical user interface (GUI) driven simulation tool is developed using MATLAB 7.2 software [12].

First two figures give you a brief outline of the polybag manufacturing process before proceeding to the core of the research work. Figure 1 shows the simplified film blowing process [1,2]. In film blowing, single screw extruder melts the polymer and pumps it into a tubular die and air is blown into the center of the extruded tube causing an expansion in the radial direction. Radial and downstream extension stops at freeze line due to crystallization. Nip rolls collect the film and seal the top of the bubble to maintain inside pressure. Figure 2 gives the simplified production process flow diagram of a poly-bag manufacturing system [11] and it has four key subsequent processes: film extrusion, printing, cutting and sealing, QC and packing. The films produced in the film blowing is printed and inspected. Then poly-bag making is carried out by cutting and sealing of the films. Quality checking and inspection end the poly-bag manufacturing process.

Next section of the paper presents the preliminary analysis. Section 3 describes the mathematical model development and Section 4 presents the system simulation. Section 5 demonstrates the simulation results followed by a discussion.

PRELIMINARY ANALYSIS

The network diagram of the selected poly-bag manufacturing system is shown in Figure 3 and it depicts the actual production channels of the selected typical polybag manufacturing company. The selected poly-bag manufacturing system is a multi-channel and multi-phase system with buffer storages in work in progress (WIP) at three places: after extrusion of film, after printing, and prior to quality checking.

Although the film extrusion process is a continuous process, there is no online feeding mechanism to the printing machine and therefore, the film manufactured is accumulated on the film store until it is fed to the printing machines depending on the urgency of orders and the availability of printing machines. In case of frequently changing orders of small quantities, printing process is a bottleneck in the poly-bag manufacturing process.

At the beginning, a SWOT analysis [13] was carried out to discover the strengths, weaknesses, opportunities, and threats faced by the system. In addition, a risk analysis and a bottleneck analysis were carried out to identify the risks faced and bottlenecks impeding in the poly-bag manufacturing system [14]. The results of the analysis were used to propose suggestions to improve the polybag manufacturing system and to identify the characteristics of the system model to be developed for the simulation. In the analysis, the layout of the factory and the bag manufacturing process were visually inspected and the required data (demand, production, timing, etc.) were gathered. Few discussions were taken place with top management, factory manager, and few experienced labours.

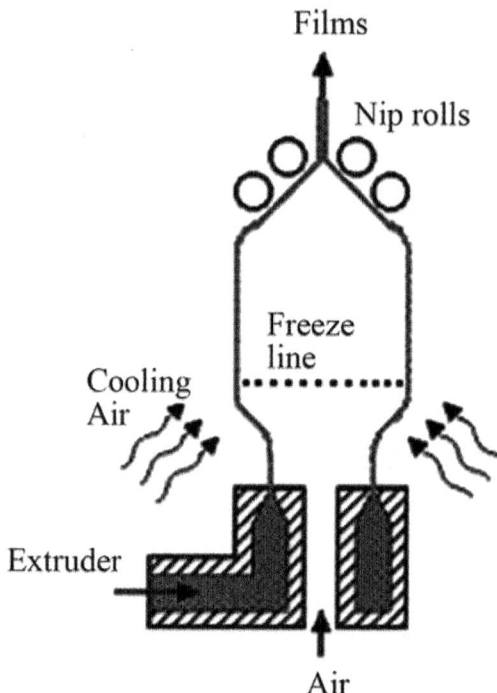

Figure 1: Simplified film blowing process.

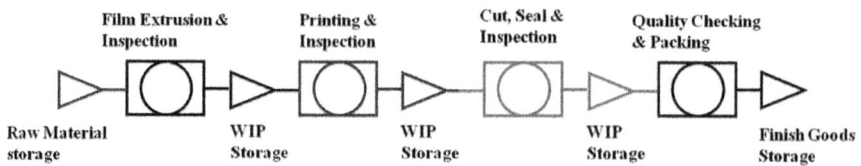

Figure 2: Simplified production process flow diagram.

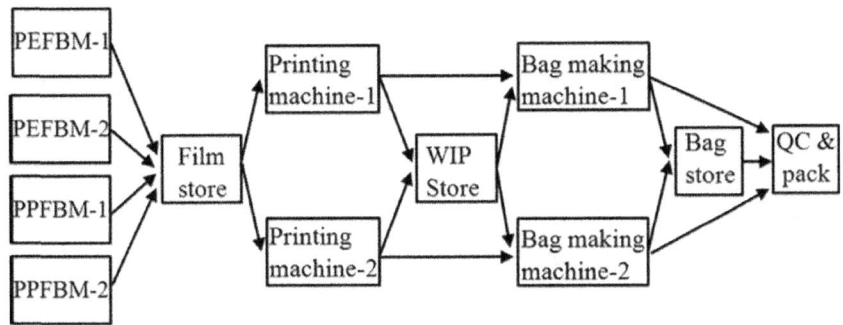

Figure 3: Network diagram of manufacturing system. PEFBM and PPFBM indicate Polyethylene Film Blowing Machine and Polypropylene Film Blowing Machine, respectively.

SWOT Analysis

The SWOT analysis was carried out and SWOT chart was drawn to indicate the strengths, weaknesses, opportunities and threats of the poly-bag manufacturing system. In the SWOT analysis, the strengths such as forward integration with member companies, availability of regular buyer base, relatively consistent number of buyers, well trained highly motivated staff, regular skill development, training and development programs, workshops, and international quality systems certification of ISO 9002 were identified. The main identified weaknesses of the system were higher wastage, improper line balancing, lack of inventory control, idling of production line, lack of record maintenance, relatively low productivity, and higher lead time. Opportunities and threats are from external origins. The opportunities for the system were identified as technical advancement of poly-bag manufacturing process, technology development of degradable poly-bags products, and boom in apparel manufacturing process creating a high demand for accessorized polybag industry. The threats from external environment to the system were legal restrictions implemented for some types of poly-bag products, environmental issues, fluctuation of raw material prices, and lower prices offered by the competitors (price competition). In addition, some technical problems such as thickness controlling of films and higher machine setting time (mainly in printing section) were the other challenges faced by the poly-bag manufacturing system. The final SWOT chart developed is shown in Figure 4. The chart shows the strengths, weaknesses, opportunities, and threats and their origins.

Risk Analysis

A risk analysis method was developed to identify the risks faced by the system. The problems that limit the manufacturing and business activities of the system were the risk of the system. If there was nothing to be done to reduce a bottleneck, the reasons of that bottleneck also become a risk. By investigating the problems that limit the activities of the system and unsolvable bottlenecksthe risks of the system could be identified. In the risk analysis, the risks affecting the manufacturing and the business of the system were identified and they were prioritized according to their impact on the system. Then weights were given from scale of 10 according to the priority of risks. The risk bar-chat was drawn from the weight values of risks to indicate the effect of risks. In the risk prioritization, the qualitative effect of each risk for the production and the business of the system were calculated under few criteria: productivity, performance, output, and profit. Since most of the data obtained were qualitative, a five-point "Likert scale" was used to find quantitative value of each criterion. The five-point "Likert scale" used is shown in Figure 5. Then the weighted averages of the quantitative values were calculated using the following equation.

$$\text{Weight} = \frac{\Sigma Q_L \times 10}{Q_{VH} \times N}$$

(1)

where Q_L, Q_{VH}, N are the quantitative value from "Likert scale", the quantitative value of "Very High" in "Likert scale" and the numbers of criteria, respectively.

The identified risks for the system can be listed in descending order of priorities, from the highest priority as, a) increasing material prices; b) legal restrictions implemented for some types of poly-bags; c) environmental issues; d) price competition; e) increase of alterative product; and f) customers switching to alternatives. Their estimated weight values were 7, 5.5, 5, 4.5, 4.5 and 3.5, respectively. Table 1 shows the quantitative values of risks. The results of risk analysis are shown in risk bar-chart of Figure 6. The chart shows the weight values and the risk priority of identified risks. a: increasing material prices, b: legal restrictions, c: environmental issues, d: price competition, e: increase of alternative products, f: customers switching to alternatives.

Bottleneck Analysis

In the bottleneck analysis [15], all the bottlenecks of the process that the system experienced were supposed to be found. All the places of the process where the actions are

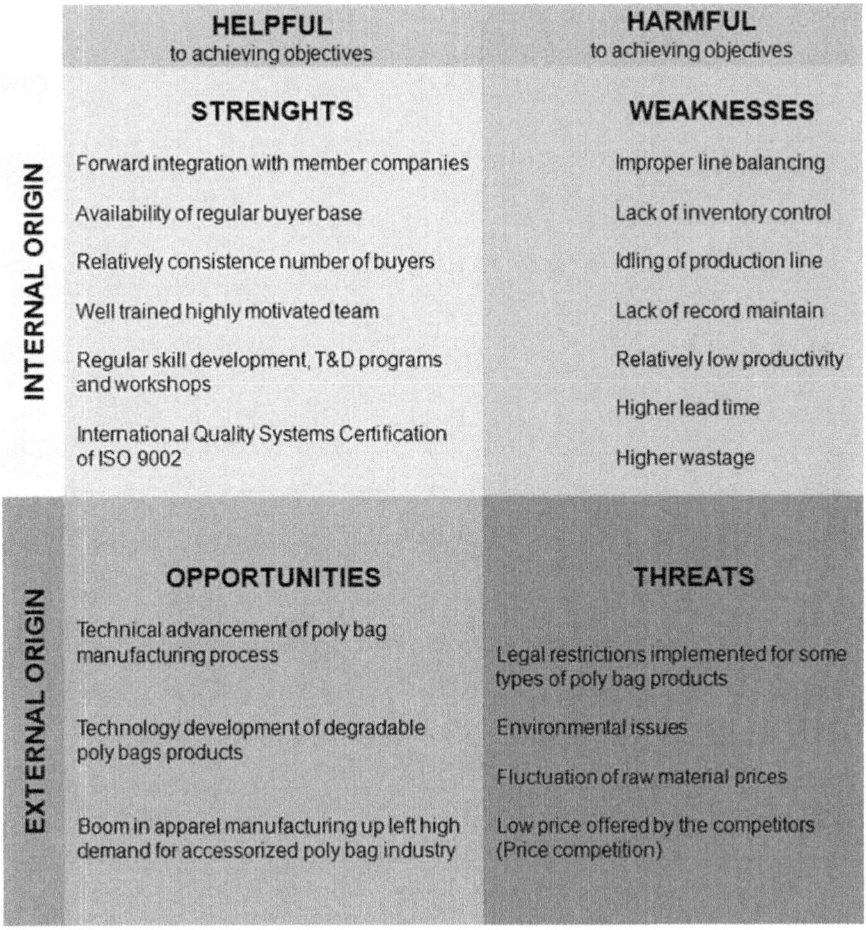

HELPFUL
to achieving objectives
HARMFUL
to achieving objectives
INTERNAL ORIGIN
STRENGHTS
Forward integration with member companies
Availability of regular buyer base
Relatively consistence number of buyers
Well trained highly motivated team
Regular skill development, T&D programs and workshops
International Quality Systems Certification of ISO 9002
WEAKNESSES
Improper line balancing
Lack of inventory control
Idling of production line
Lack of record maintain
Relatively low productivity
Higher lead time
Higher wastage
EXTERNAL ORIGIN
OPPORTUNITIES
Technical advancement of poly bag manufacturing process
Technology development of degradable poly bags products
Boom in apparel manufacturing up left high demand for accessorized poly bag industry
THREATS
Legal restrictions implemented for some types of poly bag products
Environmental issues
Fluctuation of raw material prices
Low price offered by the competitors (Price competition)

Figure 4: SWOT Chart.

01
02
03
04
05
OK
Below High
High
Above High
Very High

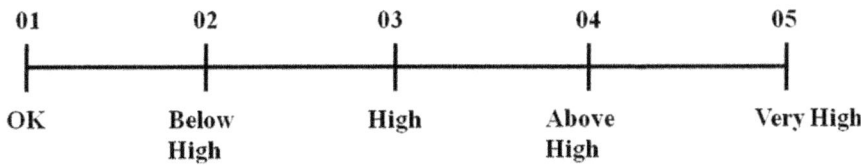

Figure 5: Five-point "Likert scale".

Table 1: Average quantitative values for each criterion

Risk	Quantitative value form Likert scale				
	Productivity	Performance	Output	Profit	Total
Increasing material prices	02	03	04	05	14
Legal restrictions implemented for some types of poly bags	03	03	02	03	11
Environmental issues	02	02	03	03	10
Price competition	01	01	03	04	09
Increase of alterative products	01	01	03	04	09
Customers switching to alternatives	01	01	02	03	07

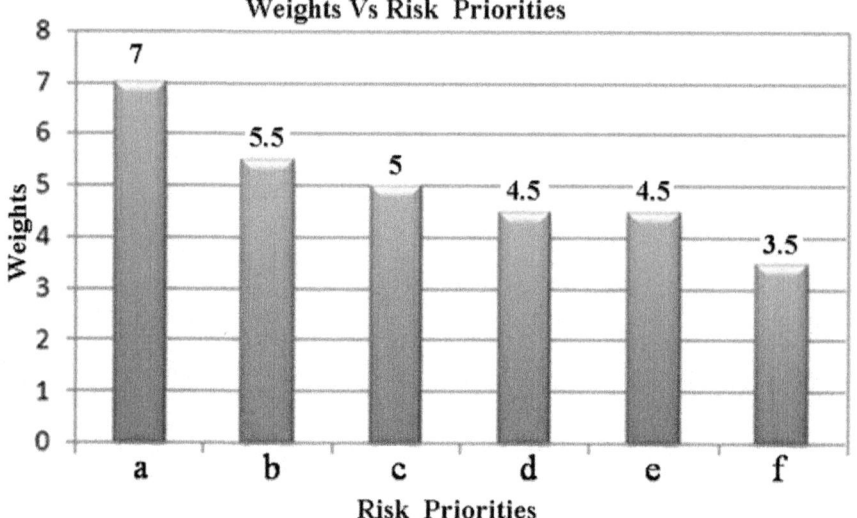

Figure 6: Risk bar-chart.

taken place were inspected and the tell-tail signs of bottlenecks such as poor response time, too long queue, insufficient resources, too slow actions, low capacity of machines etc. were examined. Then the reasons of telltails signs were further investigated. The reasons found from the investigation were the bottlenecks of the process. In the bottleneck analysis, the identified reasons for the bottlenecks were delay in quality checking, improper line balancing of two production lines, lack of inventory control especially in WIP, idling of production lines, and improper process layout.

The capacity/time vs. processing event chart was drawn for production of poly-bags to identify major bottleneck events. The capacity/time vs. processing event chart for a production of a particular polythene bag is shown in **Figure 7** as an example. The chart shows the capacity/time for main processing events. Processing events are 1: film extrusion, 2: printing, 3: quality checking, 4: handling & packing. The size of selected bag was 52 cm x 15 cm with 150 units gauge. It has one colour print. The chart shows the production capacity per unit time of major processing events. From the chart, it was identified that the QC gives the lowest capacity/time. i.e., QC is the major bottleneck of the selected poly-bag manufacturing system.

DEVELOPMENT OF THE MATHEMATICAL MODEL

Simulation of the system was carried out to study the system and further identify the problem of the system. For the simulation, a mathematical model was developed through a careful study of the manufacturing system. The objectives of modeling and simulation were the optimization of the production time for a given type of a polybag under available plant capabilities and to minimize the material wastage under current production setup. In addition, the production is scheduled for the system to achieve the maximum productivity. In a multi-phase production system, the total time to produce a given item is the summation of the time required at individual phase and the time duration required in switching between the phases if applicable. In the poly-bag manufacturing system, total time to produce a given type of a poly-bag is the summation of times of film extrusion, film storage, printing, WIP storage, bag making, QC, and setup at each operation. Since the setup time is for a large number of bags, unit setup time can be neglected as compared to the time concedes for other operations. Sub-model of each event of manufacturing will be described in next subsections. In the development of sub-models, time to produce a unit length was used as the basis to accommodate different sizes of bags conveniently into the model. The final sub section is dedicated to the generation of an

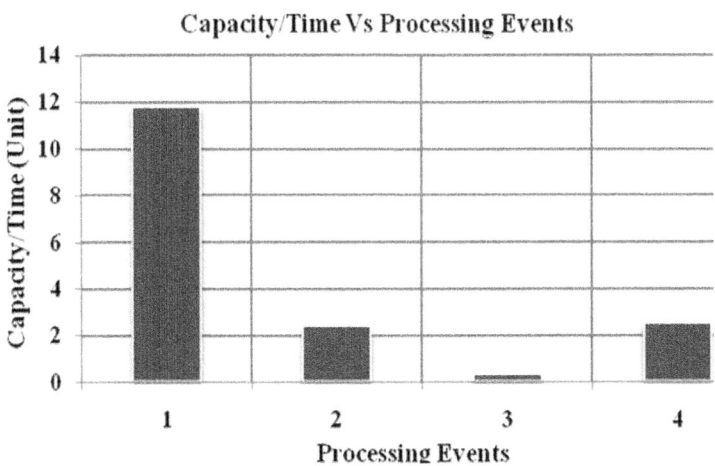

Figure 7: Capacity chart.

optimum production schedule. The objective of scheduling is to achieve the maximum productivity or find a schedule that minimizes the make-span of the jobs. Therefore, the optimum production scheduling becomes a problem of job shop scheduling. Considering the lead time and number of machines, small instant job shopping scheduling is selected for the scheduling problem.

Film Extrusion

Process of film extrusion can be modeled as;

$$T_{FE} = \begin{cases} K_a(1+W_a)L_e & \text{if the material is PE} \\ K_b(1+W_b)L_e & \text{if the material is PP} \end{cases}$$

(2)

where T_{FE}, K_a, K_b are the time to extrude the required length of a given type of bag, the film extrusion time per length of type A machine and type B machine, respectively. L_e, W_a, W_b are the film length of a given bag, the parameters representing the percentage wastes due to power failures and setting up of the film extrusion for a given product on type A machine and type B machine, respectively. These figures may vary with the frequency of power failure and the frequency of style variations. However, typical average values for medium volume production have been used for the simulation. Type A machine uses the polyethylene whereas type B machines uses the polypropylene for film extrusion. The time delay before printing is imperative and it can be given by;

$$T_m = \left[(N-1)/2\right]T_{FE}$$

(3)

where N is the number of bag lengths accumulated on a roller. Further, a delay at film store (T_{bp}) which can be set to zero in minimum time production, is also included in the model. However, delay time just before making bags (T_{bbm}) as well as the storage time of bags and delay time before quality checking (T_{bqc}) can be made zero under optimal production with proper line balancing.

Printing

Sub-model for printing is can be gives as follows.

$$T_{PR} = \begin{cases} P_a L_p & \text{one colour} \\ P_b L_p & \text{two colour} \end{cases}$$

(4)

where T_{PR}, P_a, and P_b are the time to print the required length of a given type of bag, the printing time per length on one colour machine and two colour machine, respectively. Printing time per unit length will vary depending on the nature of the print, number of colours, and the roller size of which perimeter equals to the printing length. L_p is the print length of a given bag.

Bag Making

Bag making process consists of two operations namely the sealing and cutting. The printed film is fed to a bag making machine continuously and momentarily stops it at the correct position with the aid of optical sensors for simultaneous sealing and cutting operations. The bag making time can be calculated as follows.

$$T_{BM} = F_{bm}L_b + T_{S\&C} \tag{5}$$

where T_{BM} is time taken to make the bag in the bag making section for a given product. F_{bm}, L_b, $T_{S\&C}$ are the average feeding rate of the machine including acceleration and deceleration sessions, length or width of the bag which is subjected to feeding, and cutting and sealing time of the machine respectively. Cutting and sealing time is almost a constant for a particular gauge of a film. Only at setting up of a bag making machine, few bags are subjected to quality checking for sealing strength and this operation stops once the bag making machine is set to the required strength. Only few bags are wasted in this setting up process of a large volume of production and time taken for this process can be neglected.

Quality Checking

With a time study, it was revealed that QC time of the bag is proportional to the area. Additionally, extra time is required when a defective bag is to be removed from the bundle. It is also proportional to the area of the bag since the difficulty in handling increases as the size of the bag increases. The average time for quality checking can be written as;

$$T_{QC} = Z_e B + Z_d B\alpha \tag{6}$$

where T_{QC} is the time for quality checking of a given product. Z_e, Z_d, B, α are the time needed to check a unit area of a bag for quality, additional time required to carry out the removing operation for a defective item of unit area, the area of a given bag, and the rejection ratio. The bags survived at quality checking are subsequently undergone packing operation. The values of Z_e and Z_d were found out from a time study.

Handling and Packing

Actual handling time of a poly-bag in packing operation has a non-linear relationship with the area of the bag and it depends on some other factors such as the static charges, material, and the thickness of bags. Since Sri Lanka is a tropical country with a higher humidity level, the effect of static charges can be neglected. Further, for a particular type of poly-bag, material and thickness of the bag are constant. For simplicity, counting time in handling and packing is assumed to be proportional to the area of the bag. Packing and unpacking time (T_H) which are constants, are included in handling time. Further recording of quality on the given sheets also needs a constant time (T_R).

$$T_{H\&PK} = \begin{cases} C_a B_a n + T_H + T_R \\ C_b B_b n + T_H + T_R \end{cases}$$

(7)

where $T_{H\&PK}$ is the handling and packing time for a given product. C_a, C_b, and n are the constants of proportionality in handling and packing times of polythene and polypropylene bags, and number of bag per pack respectively. T_H, T_R are handling time and report time per bag respectively.

In case of considering the minimum time to manufacture a poly-bag, storage times in between two operations should be equal to zero.

$$T_{min.manufacturing} = T_{FE} + T_{PR} + T_{BM} + T_m$$

(8)

When manufacturing 100 poly-bag items, α numbers of bags are found to be defective. Therefore, in order to survive 100 poly-bags after quality checking, it is required to manufacture $10000/(100 - \alpha)$ poly-bags. Since effective bags after QC are packed and therefore handling and packing time is only applicable to effective items. Further, delay times in the process do not get affected by the number of items in short run since it is a multi-phase feeding system. Therefore, effective through put time requirement to pack an effective item is given by

$$T_{throughput} = \left[\left(T_{FE} + T_{PR} + T_{BM} + T_m + T_{QC} \right) / \left(1 - \alpha/100 \right) \right]$$
$$+ T_{H\&PK} + T_{bp} + T_{bbm} + T_{bqc}$$

(9)

In case of considering minimum throughput time, storage times in between two operations should be set to zero. Therefore the minimum time can be expressed as follows.

$$T_{min.time} = \left[\left(T_{FE} + T_{PR} + T_{BM} + T_m + T_{QC} \right) / \left(1 - \alpha/100 \right) \right] + T_{H\&PK}$$

(10)

The rate of film extrusion, printing, and bag making are assumed to be constant for a given product on a particular type of machine. It was assumed that there is no idling time for printing and bag making machines since WIP is adequately available in stocks, initial storage of bags can entirely fulfill the demand of QC section in all the time, QC is carried out by the skilled labours and has an almost constant rejection ratio for the production, and impact of absenteeism is negligible.

Production Scheduling

Each job is characterized by a fixed order of operations, each of which is to be processed on a specific machine for a specified duration. Each machine can process at most one job at a time and once a job initiates processing on a given machine, it must complete processing uninterrupted. The objective of scheduling is to achieve maximum productivity. In other words, it is required to find a schedule that minimizes the make-span of the jobs with reasonable assumptions. Therefore, the optimum production scheduling becomes the problem of job shop scheduling. Since the system under investigation has normally 5, 6 days of lead time and number of machines used for production is below 15, the optimum production scheduling can be considered as a small instant job shopping scheduling problem. In addition, it receives few orders per day. Rather than satisfying with a good solution, it is prudent to go for an optimum scheduling under prevailing circumstances. Therefore, mixed integer programming solution method was applied. Here, it is assumed that there are N jobs and M machines, and each job follows a predetermined route, operation (i,j): processing of job j on machine i, processing time is $P_{i,j}$, jobs do not recirculate, and $t_{i,j}$ is start time of job j on machine i, $i = 1, 2, \cdots, m$ and $j = 1, 2, \cdots, n$. Ultimate aim is to minimize mixed space shop finish time, C_{max} using the mixed integer programming formulation.

$$C_{max} \geq t_{x,y} + P_{x,y} \tag{11}$$

$$C_{max} \geq t_{7,y} + P_{7,y} \tag{12}$$

$$C_{max} \geq t_{8,y} + P_{8,y} \tag{13}$$

where y is the last job received and 7, 8 are the cutting machines which end the process.

$$t_{i,j} + P_{i,j} \leq t_{i+1,j} \text{ i, j for all existing job} \tag{14}$$

$$t_{i,j} + P_{i,j} \leq t_{i,j+1} + m\left(1 - x_{i,j,j+i}\right) \tag{15}$$

$$t_{i,j+1} + P_{i,j+1} \leq t_{i,j} + mx_{i,j,j+i} \tag{16}$$

where $t_{i,j} \geq 0$, x_{ijk} EUR $\{0,1\}$.

SYSTEM SIMULATION

The mathematical model described in Equations (9) and (10) can be used to find the optimum throughput time of a given type of a poly-bag under available plant capabilities and to minimize the material wastage under current production setup. The simulation of the system can be basically divided into two categories namely system simulation and Monte Carlo simulation. System simulation is employed in deterministic processes and the performance of the system depends on parameters of the systems and the operational algorithms used. Selection of the best operational strategy and the optimal tuning of the system parameters can be done by means of system simulation. If the operations of the system mimic stochastic nature such as arrival of customer orders that are beyond the control of the system, and such system outputs can be obtained with Monte Carlo simulation. Randomness in the simulation process is achieved by the generation of random numbers and random observations are generated through inverse transformation method with an appropriate probability distribution. In generation of corresponding probability distribution, Monte Carlo sampling process is used. The production system under investigation is a hybrid system of deterministic nature and stochastic nature. Therefore, both simulation techniques were jointly used in obtaining the simulated results.

All the time durations at each production stage, QC, packing, and delay of film blowing on the machine can be calculated based on the real factory data. However, T_{bp}, T_{bbm}, and T_{bqc} depend on the urgency of the order, production quantity, similar productions in processing, and operational decisions of the management. Therefore, it mimics uncertainty and the process is of stochastic nature. Hence Monte Carlo sampling process can be devised in stochastic simulation model. The real data obtained for time-before-printing, time-before-bag-making, and time-before-QC are used to calculate the probabilities from random observations using inverse transformation method. Since the delay time is a continuous variable, data are divided into several classes in applying inverse transformation method in the above three cases.

Table 2: Delay time before printing (T_{bp})

Delay time classes (days)	Probability $P(x)$	Cumulative $F(x)$	Random variable range ($r1$)
0	0.30	0.30	00 - 29
1	0.25	0.55	30 - 54
2	0.20	0.75	55 - 74
3	0.15	0.90	75 - 89
4	0.10	1.00	90 - 99

Table 3: Delay time before QC (T_{bqc})

Delay time classes (days)	Probability P(x)	Cumulative F(x)	Random variable range (r2)
0	0.24	0.24	00 - 23
1	0.32	0.56	24 - 55
2	0.21	0.77	56 - 76
3	0.11	0.88	77 - 87
4	0.08	0.96	88 - 95
5	0.04	1.00	96 - 99

Table 4: Delay time before bag making (T_{bbm})

Delay time classes (days)	Probability P(x)	Cumulative F(x)	Random variable range (r3)
0	0.05	0.05	00 - 04
1	0.15	0.20	05 - 19
2	0.25	0.45	20 - 44
3	0.20	0.65	45 - 64
4	0.15	0.80	65 - 79
5	0.10	0.90	80 - 89
6	0.06	0.96	90 - 95
7	0.04	1.00	96 - 99

Probabilistic tables obtained for various delay times are given in Tables 2-4.

SIMULATION RESULTS AND DISCUSSION

Computer simulation was carried out using MATLAB 7.2 software package. Table 5 shows the specifications of selected poly-bags for simulation trials. In addition to these specifications, required quantities, date of receive, and the deadlines for the production are also noted on the customer orders given in Table 6. These details are used as the input for simulation trials.

Table 7 tabulates results of Monte Carlo simulation for the products A, B, C, and D. The time requirement at each process and total time are calculated for the selected products and listed in Table 8. All simulation times except bag making time (T_{BM}) in Table 8 are in minutes. Bag making time (T_{BM}) in Table 8 is in seconds. Simulated processing times (T_{FE}, T_{PR}, T_{QC} and $T_{H\&PK}$) given in the Table 8 for each product are graphically shown in Figure 8. Minimum manufacturing time for each product is given in Figure 9. The generated production schedule for the customer orders of Table 6 is given in Table 9.

The menu hierarchy of the developed simulation tool, PolySim is shown in Figure 10. The simulation tool is driven by GUI and it can be used to input, edit, and process all the necessary data for the simulation. It gives the simulation output in numerically and graphically. In addition, it can generate the optimum production schedule for a given time horizon. The simulation tool includes four menu items in the menu bar: input, output, simulation, and help as shown in Figure 11. In the input menu four sub-menus are available: Load parameter, edit parameter, customer order, and exit. Load parameter and edit updating the input parameter of the simulation. Customer orders can be checked and updated from the customer orders sub-menu. In the simulation menu three sub-menus are available (namely, System Simulation, Monte Carlo Simulation, and Production Schedule) to perform system simulation and Monte Carlo simulation and to generate optimum production schedule. Output menu includes four sub-menus to display the results of system simulation, Monte Carlo simulation, Monte Carlo and system simulation, and production schedule. Help menu consists of PolySim Help and About PolySim sub-menus to provide help and also the details of the simulation tool. Snap shots of Load Parameter window and Results-Monte Carlo simulation are shown in Figure 12 and Figure 13 respectively. The GUI in Figure 12 is used to input or retrieve machine, product, and production parameters whereas the GUI in Figure 13 shows the results of Monte Carlo simulation in numerically and graphically.

CONCLUSIONS

This paper presented an analysis, a modeling and a simulation method of a poly-bag manufacturing system. In the preliminary analysis, a risk analysis and a bottleneck analysis were carried out to identify the risks and the bottlenecks of the selected poly-bag manufacturing system. In the risk analysis, risk prioritization method was proposed. A graphical user interface driven simulation tool was also developed for the simulation. In the

Table 5: Product specifications

Product	Material	No. of colours	Bag width (cm)	Bag length (cm)	Gauge of bag
A	PE	1	15	52	150
B	PE	1	18	25	150
C	PP	2	15	16	120
D	PE	2	15	24	120

Table 6: Details of customer orders

Order	Product	Quantity (bags)	Date of received
1	A	2000	13-10-2011
2	B	1200	13-10-2011
3	C	750	13-10-2011
4	D	1500	15-10-2011

Table 7: Monte Carlo simulation outputs

Product	T_{bp}	T_{bbm}	T_{qc}	$T_{min.time}$	$T_{throughput}$
1	1440	4320	1440	49.28	7249.28
2	1440	7200	0	45.40	8685.40
3	0	4320	4320	57.30	8697.30
4	0	1440	2880	81.80	4401.80

Table 8: Simulated processing times and minimum manufacturing time

Product	T_{PE}	T_{PR}	$T_{BM(S)}$	T_{QC}	$T_{H\&PK}$	T_m	$T_{min.manufacturing}$
A	0.08495	0.41600	0.90400	3.04621	0.40382	45.31	45.82602
B	0.03905	0.20000	0.8500	1.51328	0.31759	43.31	43.56322
C	0.02997	0.20000	0.75000	1.51328	0.26271	55.28	55.52248
D	0.03750	0.59259	0.73200	0.95700	0.29407	79.91	80.55229

Table 9: Production schedule (all times are in minutes)

Order	Production start time	Film blowing time	Printing time	Bag making time	QC and handling time	Throughput time	Bags print in machine	
							one colour	two colour
1	0	84.96	416.00	15.07	460.00	976.03	1000	1000
2	977.00	23.43	120.00	8.50	146.47	298.40	600	600
3	1276.00	11.24	150.00	4.69	88.80	254.73	0	750
4	1531.00	28.13	888.89	9.15	125.11	1051.27	0	1500

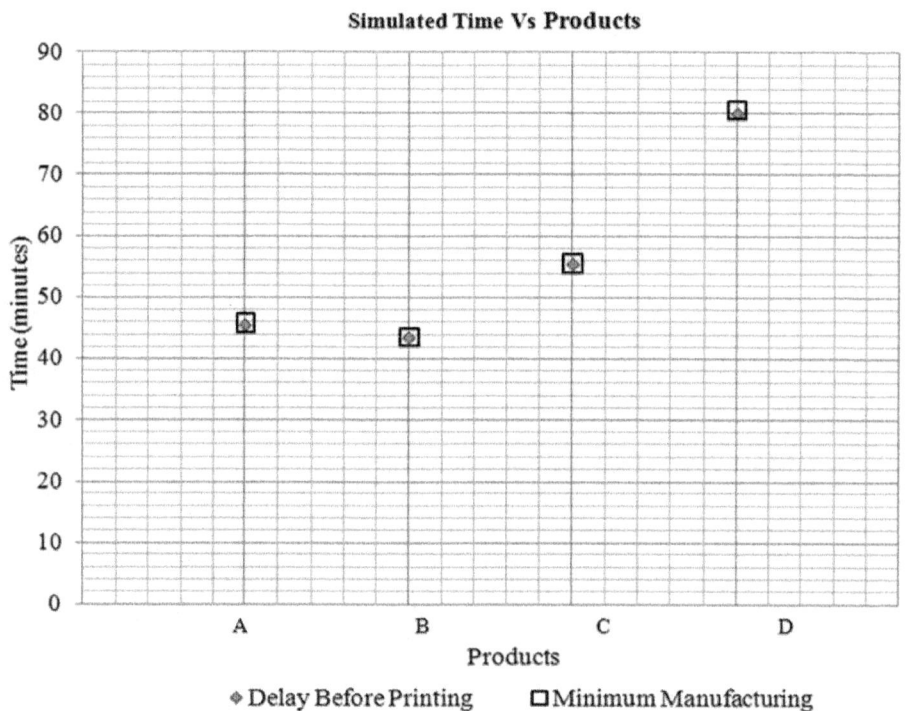

Figure 8: Simulated processing times.

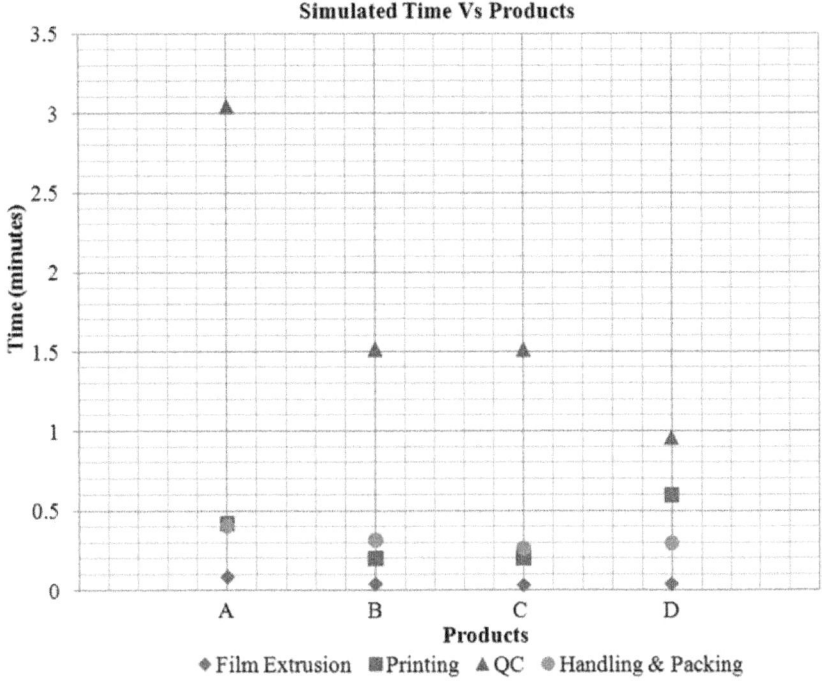

Figure 9: Minimum manufacturing time

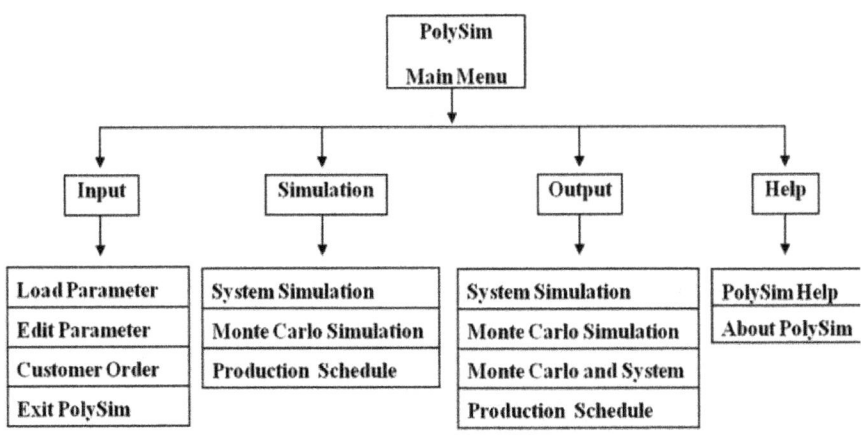

Figure 10: Menu hierarchy of the simulation tool

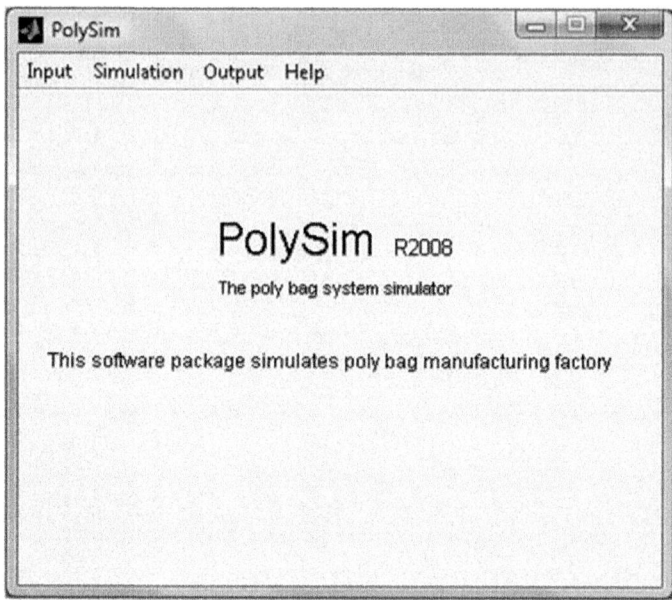

Figure 11: Snap shot of main window of simulation tool

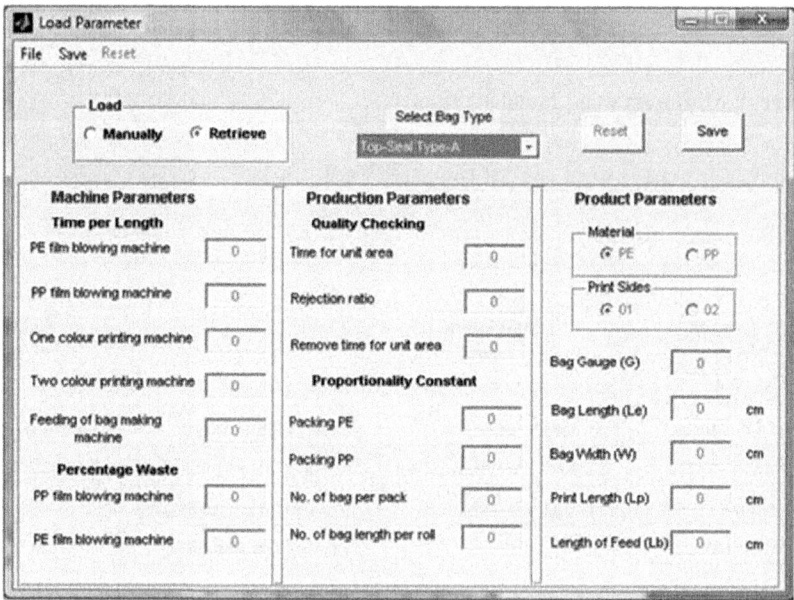

Figure 12: Snap shot of "Load Parameter" window

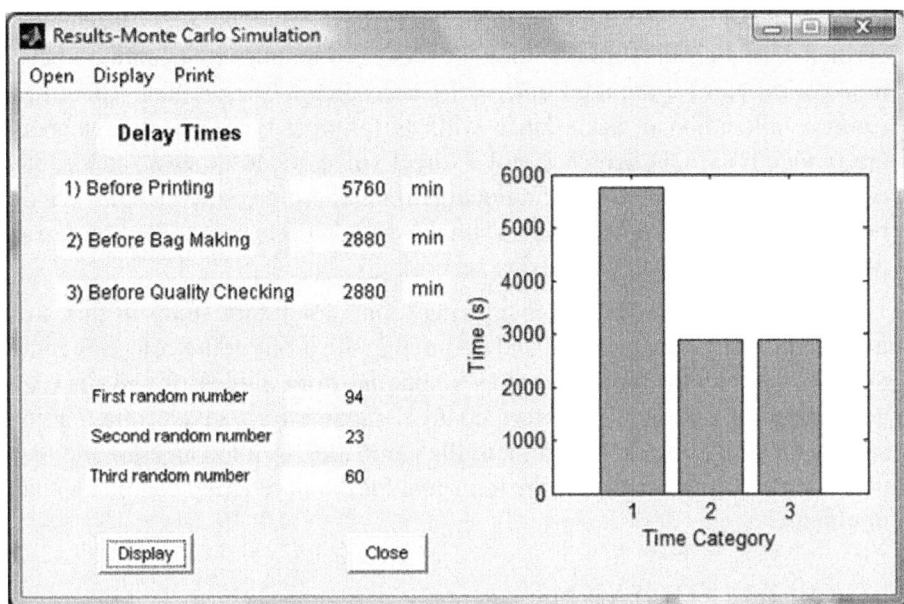

Figure 13: Snap shot of "Results-Monte Carlo Simulation" window

simulation, system simulation and Monte Carlo simulation were performed to find the optimum throughput time of a given type of a poly-bag under available plant capabilities and to minimize the material wastage under current production setup. In addition, a method of calculating optimum production schedule of the system was incorporated in the simulation tool.

From the simulation results, it is identified that the maximum contribution for minimum throughput time is from the film accumulation time on the roller. That is varying about 90% - 97% according to the product. Instead of that generally, QC time, handling and packing time, and printing time contribute about 2% - 6%, 0.30% - 0.8%, and 0.3% - 2%, respectively for the minimum throughput time in almost all of products that were used for the simulation trials. Further, it can be seen that the contribution of each process time for the minimum throughput time varies according to the product type. The contributions of the printing time, film blowing time, and bag making time for the minimum manufacturing time are about 79% - 94%, 3% - 16%, and 2% - 6%, respectively.

According to the results of the preliminary analysis and the simulation following recommendation are given. Proper production planning can be carried out to eliminate the line balancing and idling problems. Proper inventory controlling and recording method should be applied to improve the inventory control to face the challenges in periodic fluctuations in material prices though the randomness component of the material price fluctuation cannot be predicted. In the simulation, it could be identified that the contribution of each time component for the total time varies according to the product. The proper resource allocation in accordance with the product type causes to increase the productivity. QC, which is not a direct value added process, conceded a considerably large amount of production time. Therefore, a work study should be carried out in QC process and a suitable training program should be carried out for the staff of the QC division accordingly.

Following points will be interesting topics for future study in this area: incorporating a pricing system and optimizing the profit of the company rather than the production times, develop a material price forecasting system and generating the optimum inventory control schedule for a given time horizon, adaptation of wastage recycling into the simulation tool to calculate the level of corona treatment such that the total cost for the production can be kept at a minimum.

ACKNOWLEDGEMENTS

The authors would greatly appreciate the support of Thermo Plastics (Pvt) Ltd, Hemantha Munasinghe, Dharma Bandula, and Asiri Amarasene on this research.

REFERENCES

1.   Thomas and E. D. Butler, "Film Extrusion Manual: Process, Materials, Properties," 2nd Edition, TAPPI Press, Georgia, 2005.

2.   X. L. Luo and R. I. Tanner, "A Computer Study of Film Blowing," Polymer Engineering Science, Vol. 25, No. 10, 2005, pp. 620-629.

3.   A. Muslet and M. R. Kamal, "Computer Simulation of the Film Blowing Process Incorporating Crystallization and Viscoelasticity," Journal of Rheology, Vol. 48, No. 3, 2004, pp. 525-534. doi:10.1122/1.1718500

4.   V. Sidiropoulos, J. J. Tian and J. Vlachopoulos, "Computer Simulation of Film Blowing," Journal of Plastic Film and Sheeting, Vol. 12, No. 2, 1996. pp. 107-129.

5.   S. Brown, F. Chance, J. W. Fowler and J. Robinson, "A Centralized Approach to Factory Simulation," Future Fab International, 1997.

6.   M. Graul, F. Boydstun, M. Harris, R. Mayer and O. Bagaturova, "Integrated Framework for Modeling and Simulation of Complex Production Systems," Knowledge Based Systems, Inc., Texas, 2003.

7.   M. Schumann, E. Bluemel, T. Schulze, S. Strassburger and K. C. Ritter,

"Using HLA for Factory Simulation," Fall Simulation Interoperability Workshop, 1998.

8. J. O. Henriksen, "An Introduction to SLX," WSC'95 Proceedings of the 27th conference on Winter Simulation, 1997, pp. 559-566.

9. K. C. Ritter, "Skopeo-Animation," Accessed on 5 September 2011. http://simos2.cs.uni-magdeburg.de/skopeo/

10. http://www.thermosl.com

11. http://www.mathworks.com/products/

12. G. Houben, K. Lenie and K. Vanhoof, "A KnowledgeBased SWOT-Analysis System as an Instrument for Strategic Planning in Small and Medium Sized Enterprises," Decision Support Systems, Vol. 26, No. 2, 1999, pp. 125- 135. doi:10.1016/S0167-9236(99)00024-X

13. R. A. R. C. Gopura and T. S. S. Jayawardene, "A Study on a Poly-Bag Manufacturing System: Preliminary Analysis and Simulation," Proceedings of International Conference on Industrial and Information Systems, 28-31 December 2009, pp. 546-551.

14. J. Wang, "Process Bottleneck Analysis and Production Scheduling of Process Industry," Ph.D. Dissertation, Tsinghua University, Beijing, 2008.

Chapter 4

THE EFFICIENCY OF VARIANCE REDUCTION IN MANUFACTURING AND SERVICE SYSTEMS: THE COMPARISON OF THE CONTROL VARIATES AND STRATIFIED SAMPLING

Ergün Eraslan and Berna Dengiz

Department of Industrial Engineering, Baskent University, Eskisehir Road 22.km, 06590 Ankara, Turkey

ABSTRACT

There has been a great interest in the use of variance reduction techniques (VRTs) in simulation output analysis for the purpose of improving accuracy when the performance measurements of complex production and service systems are estimated. Therefore, a simulation output analysis to improve the accuracy and reliability of the output is required. The performance measurements are required to have a narrow and strong confidence interval. For a given confidence level, a smaller confidence interval is supposed to be better than the larger one. The wide of confidence interval, determined by the half length, will depend on the variance. Generally, increased replication of the simulation model appears to have been the easiest way to reduce variance but this increases the simulation costs in complex-structured and large-sized manufacturing and service systems. Thus, VRTs are used in experiments to avoid computational cost of decision-making processes for more precise results. In this study, the effect of Control Variates (CVs) and Stratified Sampling (SS) techniques in reducing variance of the performance measurements of M/M/1 and GI/G/1 queue models is investigated considering four probability distributions utilizing randomly generated parameters for arrival and service processes.

INTRODUCTION

Manufacturing systems are processing systems where raw materials are transformed into finished products through a series of workstations. It is important to find an alternative design process to obtain desired performance in a manufacturing system based on management decision. A service system is also a processing system where one or more service facilities are provided to customers, patients, and paperworks.

The use of simulation for the modeling of service and manufacturing systems has greatly increased recently in the many areas of application such as health care systems, restaurants, cafeterias, banks, and recreation centers (cinemas, theatres), and many manufacturing systems.

In systems mentioned above, the most widely used queue model is M/M/1. The queue model refers to exponential arrivals and service times with a single server and one line shown in Figure 1. M/M/1 is a good approximation for a large number of queueing systems. There are many systems we encountered in service and production fields where M/M/1 model can be used for modeling these systems such as a cashier in a supermarket and a teller in a bank.

Figure 1: One server one line queue system (M/M/1).

M/M/1 is Kendall›s notation of this queuing model. The first M represents the input process, the second M the service distribution, and 1 the number of server. The M implies an exponentially distributed interarrival and service time. The M/M/1 queue system has also unlimited population and First-in First-out (FIFO) queue discipline. On the other hand, if the distributions of arrival and service processes are not Markovian, the one server queue system is named GI/G/1.

Although there are some analytic solutions for these systems, the performance measurements of them represent steady-state behavior. Therefore, in real life applications, the simulation technique is used to compute system performance measures for any time interval. Simulation is more relevant and flexible technique to solve the problems of queue systems in manufacturing and service systems [1].

The queues are used for modeling of manufacturing systems, for example, inventory models, flow line, and JIT production systems. The unbalanced flow line capacity in the manufacturing systems can constitute a product or semiproduct queue which can be usually modeled as M/M/1 or GI/G/1. The queues can cause a bottleneck in front of the machines in the job shop (see Figure 2). The delay caused by bottlenecks in manufacturing systems increases the unit cost, decreases the productivity, and which in turn affects the competitiveness of the companies in the market negatively. This is the most common research area of Industrial Engineering in managerial and operational system analyses.

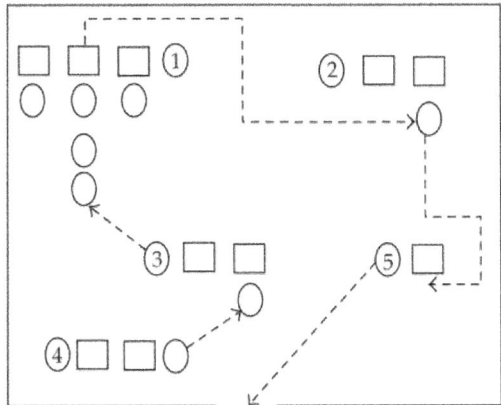

Figure 2: M/M/1 queues of the products/semiproducts in a job shop (represents a server◯ represents a manufacturer of a semiproduct).

In service systems, the success of a company, besides using the resources efficiently, depends on winning customers and keeping them. Any lost time by customers standing in the queues accounts for loss of profit and usefulness for the service companies.

To address such problems, simulation technique is used as a flexible modeling tool to investigate and solve the queue problems occuring in manufacturing and service systems. Because random samples from probability distributions are used to drive a simulation model, outputs of the simulation model are just particular realization of random variables that may have large variances. For the reasons mentioned, there has been a rapid growth of interest in the use of variance reduction techniques (VRTs) for improving the accuracy of simulation outputs. Thus, the VRTs can be used through the run of the simulation models to obtain more precise results.

Therefore, in this study, we investigated the effect of two different VRTs on the M/M/1 and GI/G/1 queue models.

The common VRTs are Common Random Variables (CRVs), Antithetic Variables (AV), Control Variates (CV), Stratified Sampling (SS), Importance Sampling (IS), Indirect Estimation (IE), and Conditional Expectation (CE) [2]. The studies on variance reduction (VR) began in the 1950s. In the years before advance computer technology, AV was used in Monte Carlo simulation. Kleijnen was interested in CRV and AV [3]. The CV technique was developed at the end of 1970s and used in a queue simulation by Carson [4], Lavenberg et al. [5], and Wilson and Pristker [6]. Law's formula was the basic of IE and Queue Theory [7]. Cartel and Ignall [8] used this formula in CRV simulation. In the following years, Nelson [9] tried the well-known VRTs in dynamic systems. Until the 1990s, there are several authors that have studied computer simulation of the VRTs, for example, Kleijnen [3], Lavenberg et al. [5], Carson and Law [10], Carter and Ignall [8], Iglehart and Shedler [11], and Wilson and Pristker [6]. In the last decade, VRTs have been used in several areas. Statistics and simulation output analyses are the primary fields and mathematics, chemistry, medicine, biology, quality improvement, portfolio analysis, pricing, flexible manufacturing systems, scheduling, stochastic networks, nuclear chemistry, oceanography, and biophysics, and Markov processes follow them. These studies shortly listed below.

Dengiz et al. [12] used the AV in the stochastic networks, Nava [13] used the VRTs in comparing the simulation models, and Shih and Song [14] in regenerative simulation. Crawford and Gallwey [15] took into account the bias of computerized simulation studies, Dahl [16] in diffusion with CV technique, Vegas et al. [17] in dichotomous response variables, Kawrakow and Fippel [18] in calculating photon dose, Plante [19] in supplier interactions of the companies, Glasserman et al. [20] in estimating of risk values for the investments, Srikant and Whitt [21] in loss model simulation, Taylor and Heragu [22] in workshop flows for shortening operations time in flow shop, Cancela and El Khadiri [23] in increasing the network reliability, Moreni [24] for establishing the pricing options of the companies, Jourdain et al. [25] in polymeric fluids in engineering, and Fumera et al. [26] in bagging optimization. The VRTs have been used in Monte Carlo simulation studies in the recent years by Skowronski and Turner [27] for statistical tolerance synthesis, Pacelli and Ravaioli [28] for semiconductor devices in electronics department, Constantini [29] for reflecting diffusions, Fitzgerald et al. [30] for determining the dynamic levels of systems, and Baker and Hadjiconstantinou [31] for Bolztman equation in fluids mechanic.

The VR and queuing models being coupled were examined in only a few studies; Görg and Fuß [32] used ATMs in runtimes evaluation, Arsham [33] in calculating the score function estimation, Meles [34] in branching the optimal numbers in mathematical models, Jocabson [35] in harmonic gradient estimator, and Schmeiser and Taaffe [36] in queuing network studies. Sabuncuoglu et al. [37] used two input and two output VRTs to measure the performance of VRTs under finite simulation run lengths and analyzed their effects considering three different types of systems: M/M/1, serial production line, and (S,s) inventory control systems.

Previous research in the area mainly focuses on applications of variance reduction techniques on M/M/1 queues. While the first objective of this paper deals with comparison and analysis of the performance of both VRTs (CV, SS) on the simple queue systems such as M/M/1 and GI/G/1, we also investigate the effects of different distributions used for modeling of arrival and service processes of these models utilizing experimental design analysis.

In this study, the average waiting time (AWT) and average number of customers (ANCs) are considered as system performance measurements. CV and SS techniques are used for the variance reduction of simulation outputs. The efficiencies of each technique on the simple queue models are investigated for four different distributions. These distributions which are exponential (in this case, the queue model is called as M/M/1 using Kendall's notation), uniform, triangular, and normal (in these cases the queue model is named as GI/G/1 using Kendall's notation) are used in interarrival times and service times. The randomly selected four parameter sets for four distributions are stated for experiments. The results of factor analysis are given in detail.

This paper will proceed as follows. The next section reviews some VRTs, and experimental analyses are described in Section 3. Finally, the research results and conclusion remarks are summarized in Section 4.

THE VARIANCE REDUCTION TECHNIQUES (VRTS)

The general specifications of CV and SS techniques are reviewed below.

Control Variates (CV) Technique

The basic purpose of CV is to introduce correlation among observations so as to reduce the variance. Using "Control Variates", true estimation statistics based on a secondary estimation value, and difference between its estimation values are ascertained. With this technique, instead of direct estimation of the parameter, the possible relationship between the problem undertaken and the analytic model is considered (see (2.1)–(2.3) [6, 38, 39].

Let X_n be a series of the first 100 customer delays in queue, and let X be an output random variable representing the average of the first 100 customer delays in queue:

$$X = E[X_n]. \tag{2.1}$$

The value of X is estimated during simulation period.

Let the secondary random variable, Y, formed from independent identically distributed random variables i.e., the service times of the first 99 customers and its expectation v EY be known because service times are generated from some known input distributions mentioned in Section 1. It is obvious that larger than average service times tend to lead to longer than average delays and vice versa. Thus Y is correlated to X, positively 7. We control the output X using this relation between X and Y, where Y is the control variate..

The corrected XX$_c$ is obtained from

$$X_c = X - a(Y - v),$$

$$(2.2)$$

where a > 0 and Y >v if X$_c$ < X, a is a constant and takes the same sign with the covariance between X and Y:

$$Var(X_c) = Var[X] + a^2 Var[Y] - 2a\,Cov[X, Y].\qquad(2.3)$$

If 2a CovX, Y > a^{2}VarY inequality is valid, then X$_c$ will have less variability than X. CovX, Y is estimated through the simulation. X$_c$ is calculated using coefficient a, then a confidence interval can be built for X$_c$ for detailed information, see Law and Kelton 7. 2.2. S

Stratified Sampling (SS)

SS technique is such that the heap is divided into stratums. By converting the heaps to stratums which have smaller variances, the problems arising from sensitivity due to big variance are prevented. Here, the determination of the number of stratums is important. Increasing the stratum number results in smaller variance but decreasing the number results in loss of the estimating variance because all data cannot be accounted for in some stratums. Moreover, the more difference between the averages of heaps and stratums the more benefit is supplied. In literature, generally, it is expressed that 3–5 stratums are enough.

There are four kinds of SS available in literature. These are Common Random SS which is used in this study, Proportional SS, Appropriate Sharing Method, and Economical Sharing Method [1].

EXPERIMENTAL ANALYSIS

We consider a simple queue system which has one waiting line and one server to perform an experimental analysis. Our aim is to determine the effectiveness of CV and SS and how VRTs will avoid computational cost of simulation experiments in obtaining more precise results. The effects of four different probability distributions having randomly generated parameter(s) for arrival and service processes on the precision of simulation output are also examined.

VRTs have two levels as CV and SS. For the arrival and service processes, exponential, uniform, triangular, and normal distributions are selected. Each distribution is assigned to arrival and service processes with randomly selected parameter values as given in Table 3. Comparison and analysis are carried out using statistical output analysis for a single system. Two performance measurements, average waiting time (AWT) and average number of customer (ANC), are considered as system outputs. The reason we use a simple queue system with one line and one server (called M/M/1 with exponential distribution for arrival and service process and GI/G/1 with the other general distributions) is to increase the range of application areas in practice and also to obtain mathematical models for them [2, 7, 40].

Variance Reduction with CV

To determine the efficiency of CV on M/M/1 and GI/G/1 models, the simulation code of M/M/1 queue model is used [7]. Some necessary modifications are performed and subroutines are used for eight design points. The two levels of VRTs and the four levels of distributions are then tested. The purpose of CV technique is to combine an appropriate definition of variables, which depend on the service times. These services are run in a controlled environment. During this study, the values of X_n waiting times in queue are created and reserved in a hidden file, and then the values are used for the remaining steps of CV. The same operations are done for the service times with their mean being v. The constant a is estimated from sample depending on the calculated covariance between X and Y using 3.1. Thus, outputs of simulation model, X, are adjusted using CV.[41].

$$a^* = \frac{\text{Cov}(X, Y)}{\text{Var}(Y)}.$$

(3.1)

Variance Reduction with SS

The SS technique works under principles of separation of the heap into stratums and reflects the process of VR of each stratum itself on the overall variance. The necessary modifications are performed on the M/M/1 simulation code for stratification. The simulation model is run ten times for 5 stratums for this study, considering 100 customers for each, to obtain the sensitivity and small variance. The replications are done via randomly selected 12 seeds shown in Table 1. After the first run of each seed, a repetition is avoided by using the same initials and a second set of random numbers for the second stratum having 100 customers [1, 4].

Table 1: A sample application for 5 stratums of SS

Random seeds	Min	Max	Customer numbers per stratum					
			(0–0.5)	(0.5–1)	(1–1.5)	(1.5–2)	(2+)	Total
65000	0.0018	8.5093	28	24	17	11	20	100
70000	0.0008	8.2443	26	26	16	12	20	100
75000	0.0003	6.8270	22	22	18	11	22	100
80000	0.0011	6.9670	23	23	18	11	21	100
85000	0.0052	5.4230	24	24	18	11	19	100
90000	0.0012	6.8600	24	24	16	14	20	100
95000	0.0018	8.2440	24	24	17	11	20	100
100000	0.0045	6.9670	25	25	17	12	19	100
105000	0.0009	5.6330	23	23	19	11	19	100
110000	0.0018	8.5090	23	23	17	13	18	100
115000	0.0012	8.3485	23	23	17	12	21	100
120000	0.0012	8.2443	23	24	18	11	20	100

The five stratums are used in this study as well as the balance of number of customers. A sample application of SS is shown in Table 1. The first column shows the random seeds, the second and third columns represent the minimum and maximum values of stratums, and the next five are the frequency of them for the stratification of 100 customers. Here, one heap is converted to five stratums which have small variances. Thus, sensitivity problems based on big variances are prevented. As seen in the table, the application of this technique is difficult, as it takes a very long run time.

Computational Results of VRTs

To ascertain the effects of the techniques on the considered queuing models, F hypothesis tests are used on the variance data obtained by applying the CV and SS techniques. The hypotheses are constructed as follows:

Hypothesis

$$H_o : \sigma_1^2 = \sigma_2^2,$$

$$H_1 : \sigma_1^2 < \sigma_2^2,$$

$$F_{cal} = Y > F_{tab} = X,$$

where σ_1 and σ_2 are the variances of the outputs obtained with and without VRTs, respectively. In the confidence level of 95% $(\alpha = 0.05)$, the calculated F value (F_{cal}) is greater than F table value (F_{tab}) which means the H_o hypothesis will be rejected. Thus, the variance of the output of the simulation model the performance measurement of the system obtained with VRT is smaller. The experiments are performed with a randomly selected parameter set 2 for arrival and 1.5 for service processes. The exponential distribution is considered for F tests where $m = n = 12$.

As shown in Table 2 both of the techniques reduce variance statistically, and the CV technique is more efficient than SS for the M/M/1 queue with a randomly selected parameter set. Similar results are also obtained for GI/G/1 queue system using the uniform, triangular, and normal distributions.

Table 2: The comparison of the variances of outputs for AWT

VRT techniques	F_{cal}	F_{tab}	Results
Without CV-with CV	139	2.86	H_o: reject
Without SS-with SS	5.64	2.86	H_o: reject
With SS-with CV	24.67	2.86	H_o: reject

Table 3: The randomly selected parameter sets for four distributions

Parameter sets	Process	Exponential (β)	Uniform (a,b)	Triangular (a,b,c)	Normal (μ, σ^2)
Set 1	Arrival	1	1,2	1,2,3	0,5,1
	Service	0.5			1,1.5
Set 2	Arrival	1.5	1,3	2,3,4	0.5,1.5
	Service	1			1,2
Set 3	Arrival	2	2,4	1,3,4	1,2
	Service	1.5			0.5,2
Set 4	Arrival	2.5	2,3	1,2,4	1.5,2
	Service	2			1,2.5

The Effects of VRTs and Distributions on the Output Variance

As stated in the previous sections, two factors are considered, and the effects of these factors are investigated on the system performance measurements. Factor settings are as follows.

(1)VRTs: this factor is tested in experimental design in two levels being CV and SS.(2)The distribution type of arrival and service processes: this factor is tested in four levels: exponential (for this case queue system is called M/M/1), uniform, triangular, and normal (for these three distributions, queue systems are called GI/G/1).

The considered system performance measurements are AWT and ANC. Since this study contains two factors with two and four levels, respectively, 2 × 4 = 8 design points are required in case of full factorial design. Four replications are made for each design point, so 32 experiments are performed. The results of the experimentation are analyzed by ANOVA. The validation of ANOVA results depends on normality and independence for the error components. This is performed by MINITAB by observing the standardized residual plot graphs. The assumptions are obtained using relevant transformations to the variance data.

These operations are performed for each considered performance measurements of the queue model. The four variance values obtained from different parameter sets for AWT and ANC are stated for four distributions used with CV and SS in Tables 4 and 5, respectively.

Table 4: The variances of AWT

Levels	1.Exponential	2.Uniform	3.Triangular	4.Normal
	0.000010	0.0715000	0.0000398	0.0000155
1.Control variates	0.000780	0.0063840	0.0010400	0.0001010
	0.039400	0.0000188	0.0124800	0.0000254
	1.503000	0.0824800	0.0007170	0.0000939
	0.105	0.395	0.432	1.541
2.Stratified sampling	1.012	0.492	0.386	1.224
	0.972	1.415	0.664	1.238
	2.430	0.162	0.531	1.127

Table 5: The variances of ANC

Levels	1.Exponential	2.Uniform	3.Triangular	4.Normal
	0.0000001	0.0003459	0.0000001	0.0000002
1.Control variates	0.0000223	0.0000663	0.0000007	0.0000032
	0.0129500	0.0000000	0.0000191	0.0000003
	0.0828200	0.0001932	0.0000023	0.0000010
	0.536	0.167	0.113	5.869
2.Stratified sampling	1.592	0.537	0.045	2.664
	0.669	0.167	0.088	2.177
	0.223	0.026	0.107	1.924

The ANOVA results given in Table 6 for AWT and Table 7 for ANC provide the followings.

Table 6: The ANOVA output for AWT

General linear model: Var versus Tech; Dist					
Factor	Type levels	Values			
Tech	Fixed	2 cv ss			
Dist	Fixed	4 ex un tr no			
Analysis of variance for Var, using adjusted SS for tests					
Source	DF	Seq SS	Adj SS	Adj MS	F	P
Tech	1	4.29330	4.29330	4.29330	48.73	.000
Dist	3	0.42013	0.42013	0.14004	1.59	.218
Tech*Dist	3	0.42165	0.42165	0.14055	1.60	.217
Error	24	2.11432	2.11432	0.08810		
Total	31	7.24940				

Table 7: The ANOVA output for ANC

General linear model: Var versus Tech; Dist					
Factor	Type levels	Values			
Tech	Fixed	2 cv ss			
Dist	Fixed	4 ex un tr no			
Analysis of variance for Var, using adjusted SS for tests					
Source	DF	Seq SS	Adj SS	Adj MS	F	P
Tech	1	4.9823	4.9823	4.9823	97.65	.000
Dist	3	2.5163	2.5163	0.8388	16.44	.000
Tech*Dist	3	2.5459	2.5459	0.8486	16.63	.000
Error	24	1.2245	1.2245	0.0510		
Total	31	11.2689				

(i)   The main factor VRTs are statistically significant for AWT, others are not.(ii)The main factor VRTs, the type of distributions, and their interactions are statistically significant for ANC.(iii)Conversely, the effect of CV technique on the performance measurements is stronger than the effect of SS technique. The CV technique results in smaller variance for both considered performance measurements; the difference in VR can be easily seen $(P < .05)$..(iv)The investigation of interaction between distribution types and the VRTs show that interaction is efficient in VR technique, only for the ANC. The smallest mean belongs to the first level of the first factor, that is, CV, and the third level of the second factor, that is, triangular distribution, shown in Tables 4 and 5.

DISCUSSION AND CONCLUSIONS

Queue systems are widely used in various fields in manufacturing and the service industry. The system analysis of the queues for both industries is one of the most highly research problems in Industrial Engineering. These analyses are mainly performed by simulation technique. Simulation output analysis is used to improve the accuracy and the reliability of the performance measures of systems. For a given confidence level, a smaller confidence interval is supposed to be better than the larger one. The wide of the confidence interval will depend on variance. Generally, increased replication of the simulation model seems to be the easiest way to reduce variance but this increases the simulation costs. Therefore, VRTs are used in experiments to avoid computational cost.

In this study, the effects of CV and SS techniques were investigated for queues (with one waiting line and one service) occurring in manufacturing and service. The effects of the two factors are investigated using the experimental design analysis in reducing variance. The first factor, VRT, with two levels (CV and SS) and the second factor, distributions, with four levels (exponential, uniform, triangular, and normal) are considered with ANOVA.

The ANOVA results show that the main factor VRTs, the type of distributions, and their interactions are statistically significant for ANC. Conversely, VRTs are statistically significant for AWT; the other factors are not.

The effect of the CV technique on the performance measurements is stronger than the effect of the SS technique. The CV technique results in smaller variance for both considered performance measurements; the difference in VR can be easily seen $(P < .05)$. It can be concluded that distribution types and VRTs jointly affect variance of the ANC measure but do not affect AWT.

The smallest mean belongs to the first level of the first factor (i.e., CV) and the third level to the second factor (i.e., triangular distribution), a combination resulting in higher efficiency.

The results underline that both CV and SS VRTs reduce variance quite efficiently in the 95% confidence level. 80% of the overall variance reduction is obtained using CV technique and 43% of using SS technique.

The further results based on the design of experiment demonstrate that if the considered system is M/M/1, CV technique is efficient. If the considered model of a system is GI/G/1 and its source of randomness (arrival and service distributions) is fitted using triangular distribution, then the CV technique is preferable to obtain more beneficial results with smaller variance. The results are only valid under the current experiments for the selected two VRTs and four distributions.

It is supposed that it is more useful to extend this research considering other VRTs to investigate and solve the problems of queuing systems in the manufacturing and service systems area as future research.

REFERENCES

1. N. Iscil, Sampling Methods, Die Library, Ankara University, Ankara, Turkey, 1977.

2. B. L. Nelson, "Decomposition some well-known variance reduction techniques," Journal of Statistics and Computer Simulation, vol. 23, no. 3, pp. 183–209, 1986.

3. J. P. C. Kleijnen, "Antithetic variates, common random numbers and computation time allocation in simulation," Management Science, vol. 21, no. 10, pp. 1176–1185, 1975.

4. J. S. Carson , "Variance reduction techniques for simulated queuing process," Department of Industrial and Systems Engineering, University of Wisconsin, Madison, Wis, USA, 1978.

5. S. S. Lavenberg, T. L. Moeller, and P. D. Welch, "Statistical results on

control variates with queuing network simulation," Operational Research, vol. 30, no. 1, pp. 182–202, 1982.

6. J. R. Wilson and A. A. B. Pristker, "Variance reduction in queuing simulation using generalized concomitant variables," Journal of Statistical Computation and Simulation, vol. 19, no. 2, pp. 129–153, 1984.

7. M. Law and W. D. Kelton, Simulation Modeling and Analysis, McGraw-Hill Series in Industrial Engineering and Management Science, McGraw-Hill, New York, NY, USA, 2nd edition, 1982.

8. G. Carter and E. J. Ignall, "Virtual measure: a variance reduction technique for simulation,"Management Science, vol. 21, no. 6, pp. 607–616, 1975. ·

9. B. L. Nelson, "A perspective on variance reduction in dynamic simulation experiments,"Communications in Statistics. Simulation and Computation, vol. 16, no. 2, pp. 385–426, 1987.

10. J. S. Carson and A. M. Law, "Conservation equations and variance reduction in queueing simulations,"Operations Research, vol. 28, no. 3, part I, pp. 535–546, 1980. ·

11. D. L. Iglehart and G. S. Shedler, "Simulation output analysis for local area computer networks," Acta Informatica, vol. 21, no. 4, pp. 321–338, 1984. ·

12. B. Dengiz, A. S. Selcuk, and F. Altiparmak, "Antithetic variate in simulation of stochastic networks: experimental evaluation," in Proceedings of International AMSE Conference on Signals, Data Systems, vol. 2, pp. 41–49, AMSE Press, Calcutta, India, December 1992.

13. M. P. Nava, "On the use of variance reduction techniques when comparing simulation systems at the steady state," Computers & Industrial Engineering, vol. 29, no. 1–4, pp. 483–487, 1995.

14. N.-H. Shih and W. T. Song, "Correlation-inducing variance reduction in regenerative simulation,"Operations Research Letters, vol. 19, no. 1, pp. 17–23, 1996.·

15. J. W. Crawford and T. J. Gallwey, "Bias and variance reduction in computer simulation studies,"European Journal of Operational Research, vol. 124, no. 3, pp. 571–590, 2000.

16. F. A. Dahl, "Variance reduction for simulated diffusions using control variates extracted from state space evaluations," Applied Numerical Mathematics, vol. 43, no. 4, pp. 375–381, 2002.·

17. E. Vegas, J. del Castillo, and J. Ocaña, "Efficiency and exponential models in a variance-reduction technique for dichotomous response

variables," Journal of Statistical Planning and Inference, vol. 85, no. 1-2, pp. 61–74, 2000. ·

18. Kawrakow and M. Fippel, "Investigation of variance reduction techniques for Monte Carlo photon dose calculation using XVMC," Physics in Medicine and Biology, vol. 45, no. 8, pp. 2163–2183, 2000.

19. R. Plante, "Allocation of variance reduction targets under the influence of supplier interaction,"International Journal of Production Research, vol. 38, no. 12, pp. 2815–2827, 2000. ·

20. P. Glasserman, P. Heidelberger, and P. Shahabuddin, "Variance reduction techniques for estimating value-at-risk," Management Science, vol. 46, no. 10, pp. 1349–1364, 2000.

21. R. Srikant and W. Whitt, "Variance reduction in simulations of loss models," Operations Research, vol. 47, no. 4, pp. 509–523, 1999.

22. G. D. Taylor and S. S. Heragu, "A comparison of mean reduction versus variance reduction in processing times in flow shops," International Journal of Production Research, vol. 37, no. 9, pp. 1919–1934, 1999.

23. H. Cancela and M. El Khadiri, "The recursive variance-reduction simulation algorithm for network reliability evaluation," IEEE Transactions on Reliability, vol. 52, no. 2, pp. 207–212, 2003.

24. N. Moreni, "A variance reduction technique for American option pricing," Physica A, vol. 338, no. 1-2, pp. 292–295, 2004.

25. B. Jourdain, C. Le Bris, and T. Lelièvre, "On a variance reduction technique for micro-macro simulations of polymeric fluids," Journal of Non-Newtonian Fluid Mechanics, vol. 122, no. 1–3, pp. 91–106, 2004.

26. G. Fumera, F. Roli, and A. Serrau, "Dynamics of variance reduction in bagging and other techniques based on randomisation," in Proceedings of the 6th International Workshop on Multiple Classifier Systems (MCS ‹05), vol. 3541 of Lecture Notes in Computer Science, pp. 316–325, Seaside, Calif, USA, June 2005.

27. V. J. Skowronski and J. U. Turner, "Using Monte-Carlo variance reduction in statistical tolerance synthesis," Computer-Aided Design, vol. 29, no. 1, pp. 63–69, 1997.

28. Pacelli and U. Ravaioli, "Analysis of variance-reduction schemes for ensemble Monte Carlo simulation of semiconductor devices," Solid-State Electronics, vol. 41, no. 4, pp. 599–605, 1997.

29. Costantini, "Variance reduction by antithetic random numbers of Monte Carlo methods for unrestricted and reflecting diffusions," Mathematics and Computers in Simulation, vol. 51, no. 1-2, pp. 1–17, 1999.

30. M. Fitzgerald, P. P. Picard, and R. N. Silver, "Monte Carlo transition dynamics and variance reduction,"Journal of Statistical Physics, vol. 98, no. 1-2, pp. 321–345, 2000.

31. L. L. Baker and N. G. Hadjiconstantinou, "Variance reduction for Monte Carlo solutions of the Boltzmann equation," Physics of Fluids, vol. 17, no. 5, Article ID 051703, 4 pages, 2005.

32. C. Görg and O. Fuß, "Comparison and optimization of restart run time strategies," AEÜ-International Journal of Electronics and Communications, vol. 52, no. 3, pp. 197–204, 1998.

33. H. Arsham, "Stochastic optimization of discrete event systems simulation," Microelectronics Reliability, vol. 36, no. 10, pp. 1357–1368, 1996.

34. V. B. Meles, "Branching techniques for Markov-chain simulation (finite-state case)," Statistics, vol. 25, no. 2, pp. 159–171, 1994.

35. S. H. Jocabson, "Variance and bias reduction techniques for harmonic gradient estimator," Applied Mathematics and Computation, vol. 55, no. 2-3, pp. 153–186, 1993. ·

36. B. W. Schmeiser and M. R. Taaffe, "Time-dependent queueing network approximations as simulation external control variates," Operations Research Letters, vol. 16, no. 1, pp. 1–9, 1994.·

37. Sabuncuoglu, M. M. Fadiloglu, and S. Celik, "Variance reduction techniques: experimental comparison and analysis for single systems," IIE Transactions, vol. 40, no. 5, pp. 538–551, 2008.

38. R. Añonuevo and B. L. Nelson, "Automated estimation and variance reduction via control variates for infinite-horizon simulations,"Computers & Operations Research, vol. 15, no. 5, pp. 447–456, 1988.

39. R. Y. Rubinstein and R. Marcus, "Efficiency of multivariate control variates in Monte Carlo simulations," Operations Research, vol. 33, no. 3, pp. 661–677, 1985. ·

40. Alan and B. Pristker, Simulation and Slam II, John Wiley & Sons, New York, NY, USA, 1986.

41. A. Mohammed, D. Gross, and D. R. Miller, "Control variates models for estimating transient performance measures in repairable items systems," Management Science, vol. 38, no. 3, pp. 388–399, 1992.

Chapter 5

AN ADAPTIVE MAINTENANCE MODEL ORIENTED TO PROCESS ENVIRONMENT OF THE MANUFACTURING SYSTEMS

Xun Gong,[1,2] Yixiong Feng,[1] Hao Zheng,[1] and Jianrong Tan[1]

[1]State Key Lab of Fluid Power Transmission and Control, Zhejiang University, Hangzhou 310027, China

[2]Robotics and Microsystems Center, Soochow University, Suzhou 215006, China

ABSTRACT

We explored an adaptive maintenance model of the process environment to diagnose progressive faults in manufacturing systems. Progressive faults are usually caused by deterioration of the operating environment or aging and show stochastic properties. Many researchers have reported how to detect faults on the machine body in manufacturing systems. However, little research has been conducted on the process environment which causes progressive faults. To tackle this problem, we explored an adaptive maintenance model to detect progressive faults and repair the process environment on the E-repair location. When a difference of the environmental factor state is detected, it will combine the transcription factor and the state enzyme to locate fault source. Then the comprehensive maintenance program is derived to repair the operating environment while eliminating progressive faults. For the purpose of validation, this model was implemented on the process environment of the air separation plant. And the simulation experiments validated the feasibility and effectiveness of this method.

INTRODUCTION

Some large complex manufacturing systems are often operated under high pressures, at high temperatures, with fast material flows and complex manufacturing mechanism. In production, facility malfunction, environmental fluctuation, or feed stream instability can introduce a variety of process disturbances, which would aggravate the load of the equipment, accelerate wear and tear on the components, and increase the consumption of electricity or power. Severe combined disturbance propagations in a plant can be destructive. Obviously, such security threatening situations should be detected early, the potential impact on production should be precisely monitored, and operational solutions should be derived quickly.

Failure process in practical engineering applications mainly includes fault diagnosis and maintenance. Publishing of Beard's doctoral dissertation in 1971 marked the birth of fault diagnosis technology [1]. Since then, the fault diagnosis technology has become a research focus and scholars have conducted extensive and in-depth researchin two groups: (1) the model-based methods: some intelligent classification algorithms, such as artificial neural networks (ANNs) and support vector machines (SVM), have been successfully used for fault diagnosis of mechanical systems [2–5]. In machine condition monitoring and fault diagnosis, some researchers have used this as a tool for classification of faults [6, 7]; (2) the data-driven methods: this group of methods monitored and collected the input and output signals of the manufacturing process [8–11]. It extracted fault features from a large number of practical samples, described the relationships between faults and symptoms, and then constructed deep knowledge of expert systems [12, 13].

The concept of preventive maintenance (PM) involves the performance of maintenance activities prior to the failure of equipment [14, 15]. One of the main objectives of PM is to reduce the failure rate or failure frequency of the equipment. This strategy contributes to minimizing failure costs and machine downtime (production loss) and increasing product quality [16]. Reliability-centered maintenance emphasizes on equipment reliability and the consequences of equipment failure as the main basis for maintenance strategy [17, 18]; fault limited strategy decides whether to maintain the system or not by the failure rate and reliability as indicators [19]; condition-based maintenance strategies are monitoring the system [20, 21]; engineering systems maintenance strategy mainly considers the economic relations among the system devices [22].

The studies of environmental factors that affect product performance have focused on environmental simulation test before the operational process [23]. These methods exposed the defects of product components by the reliability enhancement testing. The adaptability of product was improved according to the scheduled test environment. The existing fault diagnosis and maintenance techniques are mostly oriented to device components or the system itself but not deep enough to the environmental factors stress which causes the failure [24].

From the year of failure statistics of the US airborne electronic equipment [25], it can be found that the fault caused by the temperature accounted for 22.2%; by the vibration accounted for 11.38%; by moisture accounted for 10%; by the dust accounted for 4.16%; by the salt spray accounted for 1.94%; by the impact accounted for 1.11%; by other causes accounted for 47.3%. From the above statistics it can be seen that the 52% of the total fault of the equipment system failures is caused by environmental stress factors of temperature, vibration, humidity, and pollution. At the same time the environmental stress factors also affect the validity of the detection data, thereby affecting the accuracy of fault diagnosis and blocking the maintenance work.

Motivated by those problems, we want to propose a maintenance method of the process environment to bridge the gap. The paper is presented as follows. Section 1 reviews briefly the development of the diagnosis and the maintenance. Section 2 presents the prerequisites of the method by analyzing the fundamental of the environmental stress response. Section 3 proposes the adaptive maintenance model to repair the abnormal process environment to be normal. Section 4 analyzes the temperature sensitivity of the air separation process (the precooling system/purification system/booster expansion turbine/ refrigeration system). Section 5presents the experiments on the air separation plant to test the feasibility and effectiveness of the adaptive maintenance model. Section 6 highlights findings of the paper and suggests potential research directions.

PREREQUISITE

Environmental stress response is defined as follows: during the P-F interval in the operating environment, it monitors the environmental factors of the equipment/system (the time from the potential failure to functional failure of the equipment called P-F interval). Once the early warning is in the potential failure state, the equipment/system is diagnosed timely to find out the disturbance source of the environmental factors. The operating environment is maintained in real time on the environmental repair location to avoid the duration of the potential failure state and the functional failure. It is shown in Figure 1.

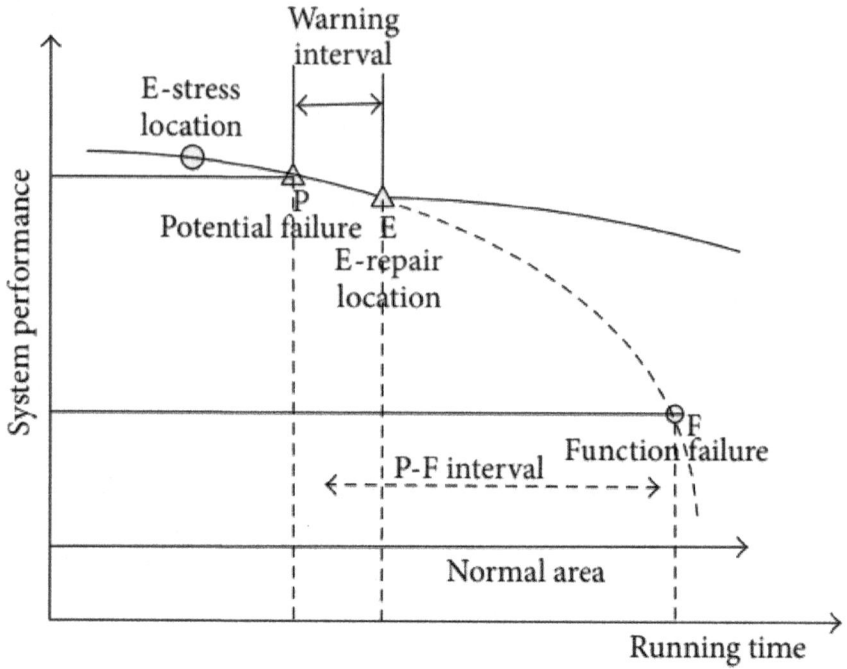

Figure 1: System performance of environmental stress response.

By the above description of the fundamental of the environmental stress response and Figure 1, it can be seen that the presence, occurrence, diagnosis, and maintenance of the operating environment potential failure have their own prerequisites. The adaptive maintenance model of the equipment/system operating environment in this paper requires some prerequisites like the following:(1)a certain degree of fault sensitivity to environmental stress;(2) determining an obvious potential failure state P;(3)less than P-F interval time length of fault warning and carrying out the environmental stress response at the environmental restoration point E;(4)the minimum P-F interval which must be long enough to arrange prevention and the environmental stress response in the potential failure process, but not the functional failure process.

The failure cumulative effect of the system is caused by environmental stress and it declines the system performance seriously. Various preventive techniques are used to diagnose potential failure timely to avoid the occurrence of functional failure. The environmental stress response on the environmental repair location repairs the operating failure environment to normal, to extend the lifetime of the system.

METHODOLOGY

Diagnostic Description

The mechanical device is constituted by the function, behavior, structure, carriers, and other design elements. In the mechanical system design theory, there are reciprocating mapping relationships among function domain, behavior domain, and carrier domain [26]. During operation, the device performance degrades because of fluctuations in E-factors. As shown in Figure 2, in order to diagnose the potential fault of the operating environment of the equipment in an abnormal operating environment, environment domain, monitoring domain, and state domain are increased beside equipment design elements domains.

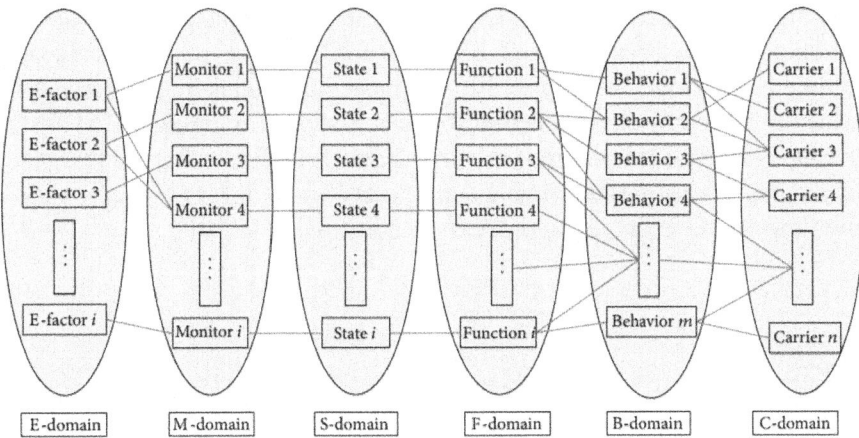

Figure 2: Domain structure mapping model.

Environmental domain is the set of E-factors in the equipment operating condition. The real-time state of the operating environment is gathered by monitoring the E-factors timely. The state data is traced back to the source of fluctuations in E-factors by diagnosing and analyzing.

The temperature is one of the E-factors which the equipment must face. Its fluctuations affect the performance of the system. For example, high temperature could cause thermal aging, structure changing, or physical expansion, while low temperature could cause material physical contraction, and temperature changes could cause the expansion and contraction, the institution stress, and so on.

Humidity is another E-factor. The high humidity stress can cause moisture accumulation and electrochemical reaction resulting in potential failures, while the low humidity stress can cause materials dried, grain, or other reactions.

The E-factors are various and random in reality, and there are mutual interactions among them. There are other E-factors leading to the system failure such as vibration/pressure/salt spray.

Adaptive Maintenance Model

The environment of the factory in which the equipment locates is extremely complex. There are a variety of disturbance sources causing the random fluctuations of the operating environment. The equipment is susceptible to progressive fault state under the E-factors stress during operation. If the abnormal operating environment is not diagnosed and maintained timely, the fault behavior is transferred and diffused among the mechanical components, and the progressive fault state also accumulates quickly. Eventually the function failure of the device would occur. Because of this, an adaptive maintenance model is proposed to diagnose and maintain progressive fault which is caused by environmental stress. It is shown in Figure 3. The adaptive maintenance model is composed of two parts. It is shown that the fluctuations of the manufacturing performance are caused by E-factors in part 1. And the monitoring and maintaining process are constructed in part 2.

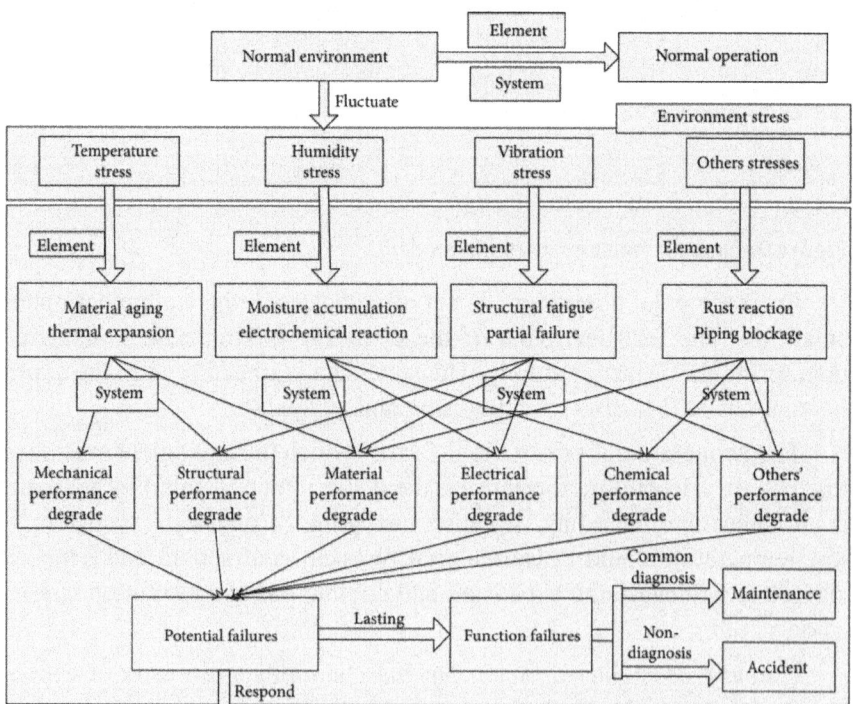

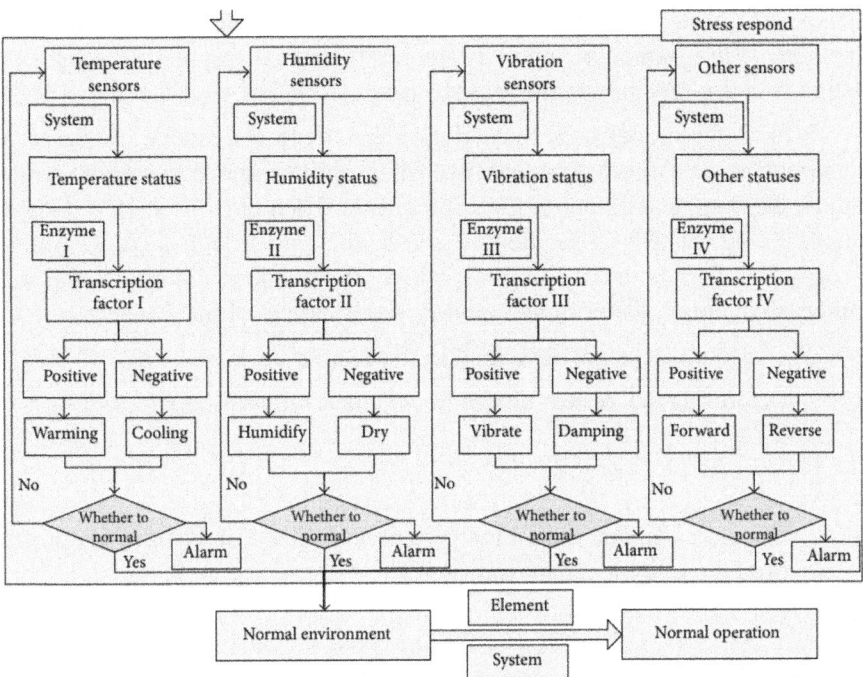

Figure 3: The adaptive maintenance model.

The random fluctuations of the operating environment are considered during the system design process. E-factors are adjusted through the adaptive maintenance model based on the progressive fault state of the system. The method keeps the system in the normal operating environment.

Firstly, the system is running in the normal environment. According to the statistics, the stochastic dynamic environmental stresses are divided into several kinds such as temperature stress, humidity stress, vibration stress, and others.

Secondly, the stresses act on various components of the system causing a variety of physical and chemical reactions and affect the performance of the whole system to the progressive fault. If the progressive fault acts constantly on the system accumulating the fault state, that would lead to the function failure of the system. In the function failure process, the traditional fault diagnosis methods can be used for diagnosis and maintenance.

Thirdly, signal monitoring and recognition technology would be applied to the state of the system environment in the progressive fault process.

Fourthly, transcription factors and the array of state enzyme are constructed by the real-time operation of the system environment state and comapped to the expert system of the fault diagnosis established by the artificial intelligence methods. Environmental repair programs to the corresponding progressive fault states are obtained in the expert system.

Fifthly, the model uses transcription factors to activate or deactivate the components of the environmental repair programs and the state enzyme to adjust the trend and extent of the components. Then the closed-loop detection conditioning system is formed to repair the dynamic E-factors. And the progressive fault is removed by responding to the online environmental stress timely to maintain the normal operating environment of the system.

The specific steps are described as follows.

(1)   Determine the system environment stress monitoring sites, L :

$$L = \left\{ l_i \mid i = 1, 2, \ldots, n \right\}.$$

(1)

(2)   The environment data of the system is obtained in the normal operation from the historical data statistics, T^0 :

$$T^0 = \left\{ t_i^0 \mid i = 1, 2, \ldots, n \right\}.$$

(2)

(3)   The range of the environment factors of the system is obtained during the normal operation through the analysis of the system performance, T' :

$$T' = \left\{ \left| t_i' \right| \mid i = 1, 2, \ldots, n \right\}.$$

(3)

(4)   The state enzyme array is constructed based on the structure and the control parameters of the components of the environmental repair programs, E. The state enzyme (e_{ij}) stands for the fluctuation extent of the failure/performance parameters of the same component of the system in different environments:

$$E = \left\{ e_{ij} = f\left(\Delta T_{ij} \right) \mid i = 1, 2, \ldots, n; \ j = 1, 2, \ldots, m \right\}.$$

(4)

(5)   The system state array is constructed by real-time monitoring of the E-factors of the operation system, T :

$$T = \left\{ t_{ij} \mid i = 1, 2, \ldots, n; \ j = 1, 2, \ldots, m \right\}.$$

(5)

(6)   The environmental stress transcription factor array is calculated by monitoring the E-factor state data, F :

$$F = \left\{ f_{ij} \mid i = 1, 2, \ldots, n; \ j = 1, 2, \ldots, m \right\},$$

$$f_{ij} = \begin{cases} 1 & \left| t'_i \right| < \left| t_{ij} - t_i^0 \right|, t_{ij} < t_i^0 \\ 0 & \left| t'_i \right| > \left| t_{ij} - t_i^0 \right| \\ -1 & \left| t'_i \right| < \left| t_{ij} - t_i^0 \right|, t_{ij} > t_i^0 \end{cases} \quad (i = 1, 2, \ldots, n).$$

(6)

Transcription factor is one of the concepts of genetics. The transcription factor array can be used in the field of fault diagnosis to determine the locations of system components which could be affected by the environmental stress and in the progressive fault state. The positive or negative of the matrix values shows out the trend of the progressive fault.

(7)   By the transcription factor array F and the state enzyme array E comapping to the expert system of the fault diagnosis, obtain the comprehensive maintenance program for the environmental stress response M. While the transcription factor value (f_{ij}) is 1, it shows that the monitoring site (l_i) is under the environmental stress and the response is positive, and while the transcription factor value (f_{ij}) is 0, it shows that the monitoring site (l_i) is not under the environmental stress and it needs no response, and while the transcription factor value (f_{ij}) is -1, it shows that the monitoring site (l_i) is under the environmental stress and the response is negative.

The repair effects of the environmental restoration program (M) is determined by comparing the environment state data of the operating system (such as T_i and T_{i+1}). If it is found out that the data and the trend do not match or exceed the regulatory range of the comprehensive maintenance (M), then the alarm would be worn. Once in this kind of situation, the components of the system should be diagnosed or the system should be upgraded to prevent functional failure happening.

SENSITIVITY ANALYSIS

The air separation plant has the typical characteristics such as electrohydraulic system coupling, the complex spatial structure, and the high failure risk. The operation reliability and the product quality of the air separation plant are sensitive to environmental stresses such as temperature, humidity, vibration, and pressure. The temperature is one of the most important environmental stresses which the air separation plant must face. Its fluctuations affect the performance of the system.

The air separation process is mainly composed of a refrigerating system and rectification system, as shown in Figure 4. The simulation model includes the compressed air system, the precooling system, the purification system, the heat exchange system, the refrigeration system, and the distillation system. And there are parts of the air separation experimental setups shown in Figure 5 (the heat exchange system).

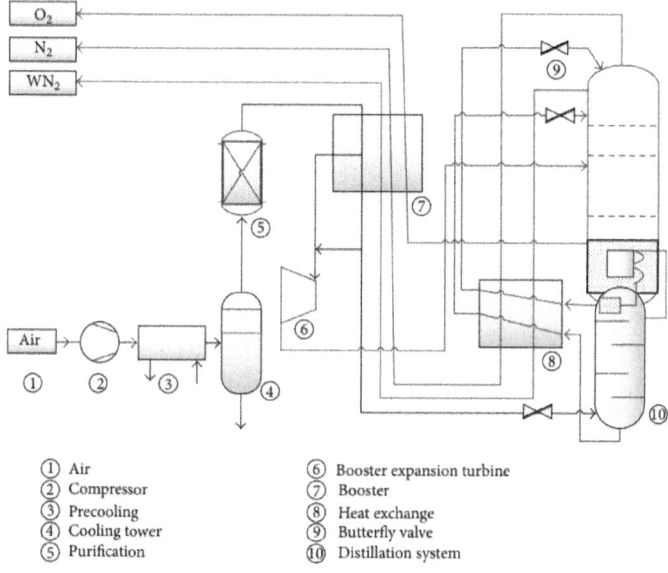

① Air
② Compressor
③ Precooling
④ Cooling tower
⑤ Purification
⑥ Booster expansion turbine
⑦ Booster
⑧ Heat exchange
⑨ Butterfly valve
⑩ Distillation system

Figure 4: The simulation model of air separation process.

Figure 5: Parts of the air separation equipment.

The precooling system of the air separation plant cools down by the circulating water. The temperature of the cooling water will decline with the drop of the atmospheric humidity and temperature and would enhance the cooling effect on the compressed air heat load; on the contrary, the temperature will rise with the rise of the atmospheric humidity and temperature and would decline the cooling effect.

The reduction of the cooling effect leads to the high temperature of the air before entering the purification system. And the high temperature air would affect the normal operation of the molecular sieve purification into the potential failure condition. Then the system production load is decreased, the productivity declines, and the power consumption increases.

85% to 90% of the cooling capacity is produced by the turbine expander of the full low pressure air separation plant. The purified air passes through the turbine expander cooling to form the raw solution. The raw solution is separated into gas products by the distillation column.

The air separation processes are simulated and the outlet temperatures of the booster expansion turbine are adjusted in the simulation. Collect the changes of the purity of the oxygen product and the nitrogen product, the oxygen extraction rate, and other data. Parts of the collected data are shown in Table 1 and Figure 6.

Table 1: Monitoring data of the booster expansion turbine

Contents	Status			
	1	2	3	4
Outlet temperature (°C)	−157.7	−161.7	−165.7	−167.7
Oxygen content in oxygen (%)	0.99953	0.99944	0.99933	0.99927
Nitrogen content in nitrogen (%)	0.9998634	0.9998636	0.9998638	0.9998639
The ratio of oxygen extraction (%)	0.6731	0.6840	0.6951	0.7007

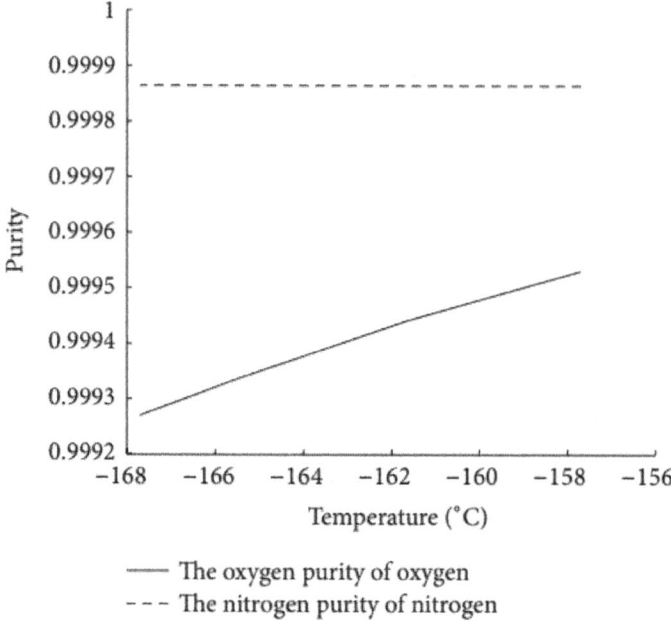

Figure 6: Diagrams of outlet temperatures of the booster expansion turbine and oxygen and nitrogen purity.

It is shown directly in Figure 6. When the temperature of the booster turbo expander declines, the expansive air is blown into the feeding plate of the upper rectifying tower and its superheat drops down. That causes the vaporization of the reflux liquid in the rectifying section of the upper tower decrease. The liquid-gas ratio in the rectifying section is higher than at the original operational temperature. The oxygen fraction condenses fully from the vapor phases to the liquid phase because of the increase of the reflux liquid flow. The oxygen content of the vapor phase drops while the nitrogen content rises.

EXPERIMENTS

A series of experiments were carried out to validate the proposed approach. The air separation process is simulated based on the data collected in the field on a certain type (7500/15000) of air separation plant in the normal operating environment. Construct the normal operating state (S_0) of the simulated air separation according to the requirements of the air separation such as the convergence and the thermal coupling of the tower systems. And then monitor the E-factors such as inlet-outlet temperatures, pressures, and flows of the key equipments such as air compressors, precooling systems, purification systems, and the booster expansion turbine.

With reference to the system historical monitoring data, the process is simulated by adjusting E-factors on the air separation system. The simulation results are shown in Table 2.

Table 2: Statistics of the air separation process compared simulation

Parameters	S	The operation in the stress state but without the environmental stress response				The operation in the stress state and with the environmental stress response			
		T_1	T_2	T_3	T_4	T_5	T_6	T_7	T_8
Air compressor									
cp_in	T	41.3	41.3	41.3	41.3	41.3	41.3	41.3	41.3
cp_out	T	72.48	75.62	78.14	79.97	72.48	75.62	78.14	79.97
Precooling system									
c_in	T	72.57	75.47	77.98	80.51	72.57	75.47	77.98	80.51
c_out	T	18.88	19.56	20.17	20.85	18.88	17.32	15.41	13.82
Purification system									
p_in	T	18.96	18.81	19.19	19.91	18.96	17.29	15.27	13.68
p_out	T	29.87	31.56	34.24	37.11	29.87	25.75	21.29	17.14
Air booster									
pb_in	T	29.87	31.44	34.12	37.03	29.89	25.77	21.35	17.28
pb_out	T	96.2	98.4	102.3	105.7	96.2	82.7	75.3	65.8
Booster expansion turbine									
e_in	T	−106.68	−104.25	−102.67	−100.18	−106.68	−108.53	−110.24	−111.08
e_out	T	−158.2	−157.5	−155.7	−153.5	−158.2	−158.9	−159.8	−161.4
Distillation tower									
Flow	O	7395	7392	7388	7382	7395	7406	7419	7430
	N	14547	14542	14539	14537	14547	14561	14572	14581
	WM	11826	11829	11831	11837	11826	11821	11814	11809
Purity	O	0.999287	0.999311	0.999338	0.999354	0.999287	0.999012	0.998772	0.998616
	N	0.918654	0.918649	0.918642	0.918636	0.918654	0.918692	0.918706	0.918712

For the operations, select the temperature stress which is one of the E-factors as the object from Table 2. Detect mainly the inlet and outlet temperatures of the air compressor, the precooling system, the purification system, and the booster expansion turbine and monitor the components of the gas products from the rectification tower. Then construct the array (L) of the system monitoring locations:

$$L = \{cp\_in, cp\_out, c\_in, c\_out, p\_in, p\_out, pb\_in,$$

$$pb\_out, e\_in, e\_out\}. \tag{7}$$

Construct the fluctuation range array (T') to maintain the healthy state based on the environment parameters of the operation system.

Consider $T' = \{40\ 60\ 60\ 4\ 4\ 6\ 6\ 30\ 10\ 3\}$. Construct the system state array (T) based on the simulated data in Table 2.

The monitoring data $(T_1 \sim T_4)$ in Table 2 are obtained by the timing simulation of the air separation process when the system is in the stress state (S_8) but without the environmental stress response. It shows that the potential failure of the system accumulates constantly

The monitoring data $(T_5 \sim T_8)$ in Table 2 are obtained by the timing simulation of the air separation process when the system is in the stress state (S_8) and the environmental stress response.

Then $T_{10,4}$ and $T'_{10,4}$ can be obtained from Table 2:

$$T_{10,4} = \begin{vmatrix} 41.30 & 41.30 & 41.31 & 41.31 \\ 72.48 & 75.62 & 78.14 & 79.97 \\ 72.57 & 75.47 & 77.98 & 80.51 \\ 18.88 & 19.56 & 20.17 & 20.85 \\ 18.96 & 18.81 & 19.19 & 19.91 \\ 29.87 & 31.56 & 34.24 & 37.11 \\ 29.87 & 31.44 & 34.12 & 37.03 \\ 96.19 & 98.41 & 102.30 & 105.70 \\ -106.68 & -104.25 & -102.67 & -100.18 \\ -158.2 & -157.5 & -155.7 & -153.5 \end{vmatrix},$$

$$T'_{10,4} = \begin{vmatrix} 41.30 & 41.30 & 41.31 & 41.31 \\ 72.48 & 75.62 & 78.14 & 79.97 \\ 72.57 & 75.47 & 77.98 & 80.51 \\ 18.88 & 17.32 & 15.41 & 13.82 \\ 18.96 & 17.29 & 15.27 & 13.68 \\ 29.87 & 25.75 & 21.29 & 17.14 \\ 29.89 & 25.77 & 21.35 & 17.28 \\ 96.19 & 82.71 & 75.34 & 65.80 \\ -106.68 & -108.53 & -110.24 & -111.08 \\ -158.18 & -158.91 & -159.83 & -161.42 \end{vmatrix}. \quad (8)$$

Locations of system components which are in the potential failure can be detected in the positive stresses by the transcription factor array (F_8). Start a negative regulation scheme for the temperature stresses to cool down or cooling the ambient system environment, while the state enzyme array (E_8) is obtained by comapping and calculating.

Make the most obvious state of the environmental stress (S_8) as the state of the potential failure which is under the temperature stress in computing of the environmental stress response. According to formula 8, the transcription factor array (F_8) can be obtained as follows:

$$F_8 = \begin{cases} 26.8 < 40 & \longrightarrow & 0 \\ 36 < 60 & \longrightarrow & 0 \\ 36 < 60 & \longrightarrow & 0 \\ 5.98 > 4 & 12.2 < 18.18 & \longrightarrow & -1 \\ 6.06 > 4 & 12.2 < 18.26 & \longrightarrow & -1 \\ 14.07 > 6 & 15.8 < 29.87 & \longrightarrow & -1 \\ 14.07 > 6 & 15.8 < 29.87 & \longrightarrow & -1 \\ 37.12 > 30 & 59.08 < 96.2 & \longrightarrow & -1 \\ 6.93 < 10 & & \longrightarrow & 0 \\ 6.9 > 3 & -165.1 < -158.2 & \longrightarrow & -1 \end{cases}$$

$$F_8 = \{0,0,0,-1,-1,-1,-1,-1,0,-1\},$$

$$E_8 = \left\{ e_{ij} = f\left(\Delta T_{ij}\right) \mid i = 1,2,\ldots,n;\ j = 1,2,\ldots,m \right\}$$

$$= \{0,0,0,1.49,1.52,2.35,2.35,1.24,0,2.3\} \tag{9}$$

Cool down the components which are under the potential failure and the ambient environment by operating the comprehensive maintenance program (M_8) to adjust the Efactors. The comparisons of the air separation process simulation between those without (solid line) and those with (dotted line) the environmental stress response are shown in Figures 7 and 8. The outlet temperatures of the precooling system, the purification system, and the air compressor are shown in Figure 7. The inlet-outlet temperatures of the booster expansion turbine are shown in Figure 8.

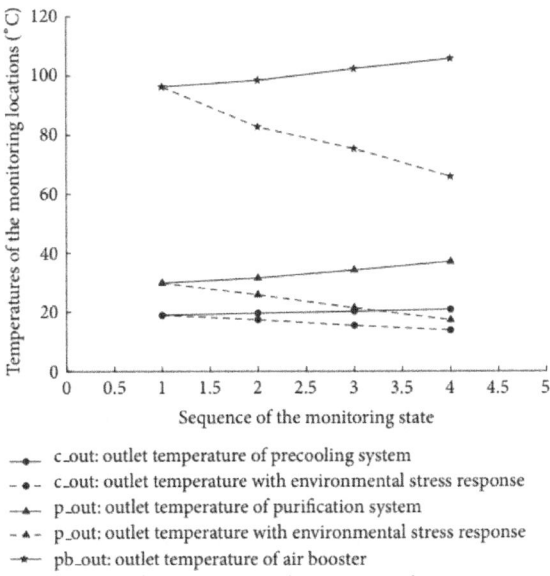

Figure 7: Contrast diagram of the running temperatures of the components.

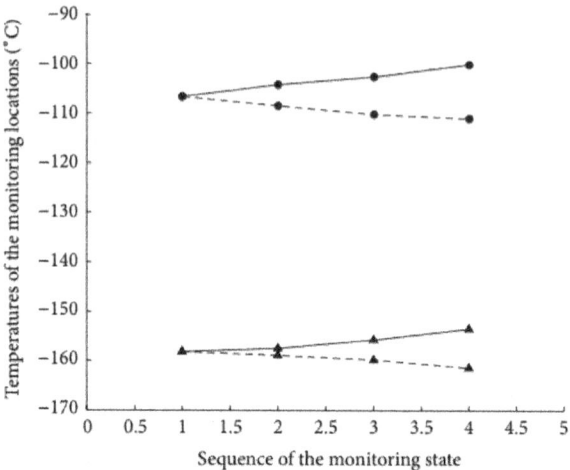

—◆— e_in: inlet temperature of the booster expansion turbine
- ◆ - e_in: inlet temperature with environmental stress response
—▲— e_out: outlet temperature of the booster expansion turbine
- ▲ - e_out: outlet temperature with environmental stress response

Figure 8: Contrast diagram of the inlet and outlet temperatures of the booster expansion turbine.

It can be seen by the comparisons that it is effective to take the environmental stress response on the operational phase of the air separation system as shown in Table 3. The reducing of the temperature means low power consumption of the manufacturing system and the increasing of production means higher benefits. Also the normal operating environment would extend the service life of the system.

Table 3: Analysis of the effect of the adaptive maintenance model

Locations	Status			
	T_4 (°C)	T_8 (°C)	ΔT (°C)	$\Delta\%$
c_out	20.85	13.82	−7.03	−33.72
p_out	37.11	17.14	−19.97	−53.81
pb_out	105.7	65.8	−39.9	−37.75
e_in	−100.18	−111.08	−10.9	−10.88
e_out	−153.5	−161.4	−7.9	−5.15
F_O	7382	7430	48	+0.65
F_N	14537	14581	44	+0.30

CONCLUSION

We presented an adaptive maintenance model to repair the process environment which caused progressive faults in the air separation plant system. This maintenance approach includes the following. The diagnostic model monitors the environmental states of the plants and also compares the inputs/outputs and presettings to detect faults. The mapping structure is constructed with the I/O environmental states and behaviors of carriers, while the state enzyme and the transcription factor array are calculated through the expert system. The comprehensive maintenance program is obtained by the comapping of the state enzyme and the transcription factor array for the environmental stress response.

For the future research, we suggest to optimize the deployment of the sensors for the model. Through preselection of sensor locations, it may improve the detection of the system with optimal cost and sensor configuration.

ACKNOWLEDGMENTS

This work was supported by the National Natural Science Foundation of China (No. 51322506 and 51175456), Zhejiang Provincial Natural Science Foundation of China (No. LR14E050003), the Fundamental Research Funds for the Central Universities, Innovation Foundation of the State Key Laboratory of Fluid Power Transmission and Control, and Zhejiang University K.P. Chao's High Technology Development Foundation. Sincere appreciation is extended to the reviewers of this paper for their helpful comments.

REFERENCES

1.   R. V. Beard, Failure Accommodation in Linear Systems through Self-Reorganization, MIT, Cambridge, Mass, USA, 1971.

2.   M. Demetgul, "Fault diagnosis on production systems with support vector machine and decision trees algorithms," The International Journal of Advanced Manufacturing Technology, vol. 67, no. 9–12, pp. 2183–2194, 2013.

3.   J. Yang, Y. Zhang, and Y. Zhu, "Intelligent fault diagnosis of rolling element bearing based on SVMs and fractal dimension," Mechanical Systems and Signal Processing, vol. 21, no. 5, pp. 2012–2024, 2007.

4.   S. F. Yuan and F. L. Chu, "Support vector machines-based fault diagnosis for turbo-pump rotor,"Mechanical Systems and Signal Processing, vol. 20, no. 4, pp. 939–952, 2006.

5.   Widodo and B. Yang, "Support vector machine in machine condition

monitoring and fault diagnosis," Mechanical Systems and Signal Processing, vol. 21, no. 6, pp. 2560–2574, 2007.

6. Widodo, B. Yang, and T. Han, "Combination of independent component analysis and support vector machines for intelligent faults diagnosis of induction motors," Expert Systems with Applications, vol. 32, no. 2, pp. 299–312, 2007.

7. Y. Yang, D. Yu, and J. Cheng, "A fault diagnosis approach for roller bearing based on IMF envelope spectrum and SVM," Measurement: Journal of the International Measurement Confederation, vol. 40, no. 9-10, pp. 943–950, 2007.

8. P. Stepanic, I. V. Latinovic, and Z. Djurovic, "A new approach to detection of defects in rolling element bearings based on statistical pattern recognition," International Journal of Advanced Manufacturing Technology, vol. 45, no. 1-2, pp. 91–100, 2009.

9. F. Pan, S. R. Qin, and L. Bo, "Development of diagnosis system for rolling bearings faults based on virtual instrument technology," Journal of Physics: Conference Series, vol. 48, article 467, 2006.

10. Angeli, "An online expert system for fault diagnosis in hydraulic systems," Expert Systems, vol. 16, no. 2, pp. 115–120, 1999.

11. Angeli and A. Chatzinikolaou, "On-line fault detection techniques for technical systems: a survey,"International Journal of Computer Science & Applications, vol. 1, no. 1, pp. 12–30, 2004.

12. B. Samanta and K. R. Al-Balushi, "Artificial neural network based fault diagnostics of rolling element bearings using time-domain features," Mechanical Systems and Signal Processing, vol. 17, no. 2, pp. 317–328, 2003.

13. T. Lindh, On the condition monitoring of induction machines [Ph.D. thesis], Lappeenranta University of Technology, 2003.

14. B. Gertsbakh, Models of Preventive Maintenance, North-Holland Publishing, Oxford, UK, 1977.

15. H. Löfsten, "Management of industrial maintenance—economic evaluation of maintenance policies,"International Journal of Operations and Production Management, vol. 19, no. 7, pp. 716–737, 1999.

16. S. Usher, A. H. Kamal, and W. H. Syed, "Cost optimal preventive maintenance and replacement scheduling," IIE Transactions, vol. 30, no. 12, pp. 1121–1128, 1998.

17. Pintelon and G. Waeyenbergh, "A practical approach to maintenance modelling," in Flexible Automation and Intelligent Manufacturing,

J. Ashayeri, W. G. Sullivan, and M. M. Ahmad, Eds., pp. 1109–1119, Begell House, New York, NY, USA, 1999.

18. G. Waeyenbergh and L. Pintelon, "A framework for maintenance concept development," International Journal of Production Economics, vol. 77, no. 3, pp. 299–313, 2002.

19. H. Wang, "A survey of maintenance policies of deteriorating systems," European Journal of Operational Research, vol. 139, no. 3, pp. 469–489, 2002.View at Zentralblatt MATH ·

20. Wiseman, "Optimizing condition based maintenance," Plant Engineering and Maintenance, vol. 23, no. 6, pp. 57–71, 2001.

21. Y. Liao, G. Lang, and L. Qu, "Precession trend analysis and balancing strategy for rotors with multi-fault," Journal of Mechanical Engineering, vol. 45, no. 8, pp. 45–51, 2009.

22. Dieulle, C. Bérenguer, A. Grall, and M. Roussignol, "Sequential condition-based maintenance scheduling for a deteriorating system," European Journal of Operational Research, vol. 150, no. 2, pp. 451–461, 2003. View at MathSciNet ·

23. Kearney, J. Marshall, and B. Newman, "Comparison of reliability enhancement tests for electronic equipment," in Proceedings of the Reliability and Maintainability Symposium, pp. 435–440, Singapore, January 2003.

24. R. Ahmad and S. Kamaruddin, "An overview of time-based and condition-based maintenance in industrial application," Computers and Industrial Engineering, vol. 63, no. 1, pp. 135–149, 2012.

25. R. D. Brillhart, D. L. Hunt, and H. Chimerine, "Multiple input excitation methods for aircraft ground vibration testing," Sound and Vibration, vol. 27, no. 1, pp. 77–85, 1993.

26. Y. Umeda, T. Takeda Tomiyama, et al., "Function behavior and structure," Application of Artificial Intelligence in Engineering, vol. 10, no. 4, pp. 177–193, 1990.

Chapter 6

THE FUNDAMENTALS OF GLOBAL OUTSOURCING FOR MANUFACTURERS

Aslı Aksoy and Nursel Öztürk

Uludag University Department of Industrial Engineering, Bursa Turkey

INTRODUCTION

Today, international competition is growing rapidly, and enterprises must always remain ahead of the competition to ensure their survival. Therefore, firms must keep pace with dynamic conditions and rapid changes, be innovative, and adapt to new systems, techniques and technologies. In a competitive market environment, customers are becoming more conscious and tend to demand a particular number of customised products at a particular speed. Furthermore, fluctuations in national economies and in the global economy create significant risks. Because of all of these factors in today's competitive environment, firms have begun to make radical changes in their management and production structures. They must also reduce costs to maintain their current position in the market. Manufacturers must be the forerunners in the competitive race in today's global markets. Today's enterprises are facing fierce competition, which is forcing them to seriously consider new applications that they can use to improve quality and to reduce cost and lead time. Manufacturers must keep pace with the dynamic requirements of the market and be receptive to reform. Because of the intense global competition among manufacturers, the supply chain must be able to respond quickly to changes, and customer–supplier relationship management is becoming increasingly important. In recent years, very few manufacturers have owned all of the activities along the supply chain. The ability to make rapid and accurate decisions within the supplier network improves the competitive advantage of manufacturers. Additionally, due to the intense global competition that exists today, firms should be reevaluate and redirect missing resources. Outsourcing plays a key role for enterprises because the cost of raw materials constitutes a significant part of the cost of the final product. Choosing the right supplier reduces purchasing costs and enhances the competitive advantage of firms. As organisations become more dependent on suppliers, the direct and indirect consequences of poor decision-making become more severe. Decisions about purchasing strategies and operations are the primary determinants of profitability. The globalisation of trade and the Internet have enlarged purchaser choice sets. Changing customer preferences require broader, more systematic and faster outsourcing decision making. An enterprise may produce a specific product itself or may outsource that specific product to achieve a production cost advantage. Global outsourcing can be defined as the forwarding of specific business to a global supplier. Global outsourcing enhances the competencies of firms while also making firm structures more flexible. In today's global markets, firms must use new methods to sustain their strength and compete. In recent years, under the influence of this intense competition, global outsourcing has become popular for firms. Firms are widely using global outsourcing to adapt to rapid changes, to reduce the effect of fluctuations, and to take advantage of know-how and current technologies. Global outsourcing allows firms to develop their core competencies and expand their flexibility. This study reviews the literature on global outsourcing. There has been a great deal of research conducted on global outsourcing within information technology (IT) and service systems. To the best of our knowledge, there have not been many studies on the global outsourcing of manufacturing or production systems. Therefore, this study focused on the global outsourcing of these systems. This paper provides a basic definition of global outsourcing and analyses global outsourcing as either an opportunity or a threat. Furthermore, this study introduces the differences between local and global outsourcing. The methods used to make global outsourcing decisions and the decision criteria used in global outsourcing are also presented in the study.

GLOBAL OUTSOURCING

Outsourcing is one of the responsibilities of purchasing departments and plays a critical role in an organisation's survival and growth. Materials sourced from outside rather than produced by in-house facilities will influence service quality and profitability (Zeng, 2000). Despite the ongoing debate over the benefits and risks of outsourcing for businesses, outsourcing has become a common approach that purchasing managers cannot ignore. Indeed, outsourcing has exhibited dramatic growth in recent years. Since the 1980s, the opinions of purchasing managers and management scholars regarding optimal firm sourcing strategies have changed significantly in two respects. First, firms have replaced vertical integration with increased outsourcing based on the conviction that lean, flexible enterprises that focus on their core competencies perform better (Quinn & Hilmer, 1994). Second, in the era of globalisation of the 1990s, enterprises were advised to use the principles of "global outsourcing" to pick the best global suppliers and thereby to improve their competitiveness (Monczka & Trent, 1991; Quinn & Hilmer, 1994). Implementing both or either of these strategies has important consequences for the structure and performance of multinational corporations. Given the rapidly shifting contours of the global economy, companies need to be able to anticipate changes in the economics and geography of outsourcing. Forward-thinking companies are making their value chains more elastic and their organisations more flexible. Furthermore, with the decline of the vertically integrated business model, outsourcing is evolving into a strategic process used to organize and fine-tune the value chain. Supplier selection and evaluation play an important role in reducing the cost and time to market while improving product quality. Supplier selection can significantly affect manufacturing costs and production lead time. Although several techniques and models have been used to select and evaluate suppliers, each technique or model has its own strengths and limitations in different situations. Therefore, it is necessary to further improve the performance and effectiveness of supplier selection and evaluation in manufacturing in different contexts. According to Boer et al. (2001), the purchasing function and purchasing decisions are becoming increasingly important. As organisations become more dependent on suppliers, the direct and indirect consequences of poor decision making are becoming more severe. In addition, several developments have further complicated purchasing decision making. The globalisation of trade and the Internet have expanded the choice sets of purchasers. Changing customer preferences requires broader and more rapid supplier selection. Outsourcing can be defined as the provision of services by an outside company when those services were previously provided by the home company. In other words, outsourcing involves focusing on a firm's core competencies while allowing services that require other competencies to be provided by other expert enterprises. Outsourcing is a strategic decision in which the buying firm attempts to establish a long-term business relationship with its suppliers (Zeng, 2000). It is not always easy to generate precise rules for the supplier selection process, but certain elements of the process remain constant. These elements may be identified based on intuition, experience, common sense, or inexplicable rules. Supplier ratings, for example, are usually generated via subjective criteria, based on personal experience and beliefs, based on the available information, and/or sometimes using techniques and algorithms intended to support the decision-making process (Albino & Garavelli, 1998). The key to enhancing the quality of decision making in the supplier selection process is to employ the powerful computer-related concepts, tools and techniques that have become available in recent years (Wei et al., 1997). In today's competitive global markets, consumers look for the highest quality products at the lowest prices, regardless of where they are produced. This trend is continuously increasing the significance of global markets and forcing enterprises to enter global markets. Furthermore, increasing pressure from foreign competitors in domestic markets is forcing companies to analyse the available alternatives as they seek to remain competitive. Monczka & Trent (1991) defined global outsourcing as the integration and coordination of procurement requirements across worldwide business units. As such, outsourcing might involve objects, processes, technologies and suppliers. Kotabe (1998) defined global outsourcing as the purchase of finished products or works-in-process from global suppliers. Under this definition, firms may purchase not only products themselves but also the services required to make these products marketable. Narasimhan et al. (2006) reported that the strategic objectives of global outsourcing are different from those of traditional purchasing. Whereas traditional purchasing focuses on minimizing procurement costs, strategic global outsourcing considers quality, delivery, responsiveness and innovativeness in addition to costs. Sourcing strategies should be incorporated into the operating strategies of buying firms to support or even improve their competitive advantage (Tam et al., 2007). Internal or global outsourcing plays an important role in firm competitiveness and growth (Zeng, 2000). Flexibility appears to be an important driver of global outsourcing strategy. Firms need to react more quickly to customer requirements, and global outsourcing is seen as a way to accomplish this. Global outsourcing may also be perceived as a way to reduce firm risk by sharing it with suppliers and simultaneously acquire the positive attributes of those suppliers (Kremic et al., 2006). The ultimate objective of global outsourcing strategy is for the firm to exploit both its own and suppliers' competitive advantage and to utilise the comparative location advantages of various countries in global competition (Kotabe & Murray, 2004).

The importance of global outsourcing has increased dramatically. Although firms may outsource for cost-related reasons, there are no guarantees that expected savings will be achieved (Kremic et al., 2006). Global outsourcing strategy requires close coordination between the research and development, manufacturing, and marketing activities of a firm. Conflicts will most likely exist between the differing objectives of these divisions. For instance, excessive product modification and development intended to satisfy a set of ever-changing customer needs will negatively affect manufacturing efficiency and increase costs. Similarly, excessive product standardization intended to lower manufacturing costs will likely yield lower customer satisfaction levels (Kotabe & Murray, 2004). Therefore, effective global outsourcing requires firms to develop a balance between effective manufacturing and flexible marketing. Global outsourcing is an expected response to competition. However, the choice of where to obtain goods and services is not an obvious decision. Rather, it is subject to continual reevaluation (Carter et al., 2008). Outsourcing strategy is an essential part of the value chain for corporate activities. Outsourcing strategy both affects and is affected by the other aspects of the supply chain (Kotabe et al., 2008). The degree of internationalization of production and sourcing is negatively related to the size of the focal country. According to Mol et al. (2005) and Buckley & Pearce (1979) when working with a sample of 156 Japanese, French, Swiss, and "Benelux" companies, found the ratios of global outsourcing to final markets to be 2.4%, 8.0%, 91.6%, and 70.7%, respectively. As Levy (2005) noted, global outsourcing is highly related to efforts to increase the organizational and technological capacity of firms. Mol et al. (2005) described global outsourcing as balancing international production cost advantage and domestic transaction cost advantage rather than characterising it as a performance-enhancing tool. The major operational problems in global outsourcing, as described by Kotabe et al. (2008), are logistics, inventory management, distance, nationalism, and a lack of working knowledge about foreign business practices. Global outsourcing has become a popular subject of study in both managerial practice and the academic literature. Conflicting results have been presented in the relevant studies. The global outsourcing strategy literature offers arguments both for and against global outsourcing strategy (Kotabe et al., 2008). According to Gottfredson et al. (2005), a recent survey of large and medium-sized companies indicates that 82% of large firms in Europe, Asia, and North America have outsourcing arrangements of some kind and that 51% use global outsourcers. However, nearly 50% say that the results achieved by their outsourcing programs have fallen short of expectations. What is more, only 10% are highly satisfied with the decreases in costs that they have achieved, and a mere 6% are highly satisfied with the results of their global outsourcing efforts overall. Mol et al. (2005) stated that global outsourcing can help a firm to enhance its competitive advantage in other markets or to improve its legitimacy. However, multinational supply chains are facing significant managerial problems related to international relations. According to Kremic et al. (2006), the expected benefits of outsourcing may include providing the same or a better service at a lower overall cost, increased flexibility and/or quality, access to the latest technology and the best talent, and the ability to refocus scarce resources on core functions. A lack of common methodology is believed to cause some outsourcing failures. Lonsdale (1999) also supported this thinking, suggesting that global outsourcing failures are not due to inherent problems with outsourcing but rather stem from a lack of guiding methodology for managers. Kremic et al. (2006) indicated that global outsourcing has potential pitfalls for strategic reasons. Gillett (1994) noted that enterprises may lose their core competences if they are not careful. If firms outsource the wrong functions, they may develop gaps in their learning or knowledge base that may hinder their ability to capitalise on future opportunities (Kremic et al., 2006). Literature also indicated that in industries with complex technologies and systems, internal synergy may decrease when some functions are outsourced. This could result in lower productivity or efficiency levels for the remaining functions (Quinn & Hilmer, 1994). Kremic et al. (2006) discussed factors that may impact global outsourcing decisions. These factors are shown below:

- Core competences: "Core competences" can be described as a strategic factor that firms use to sustain competitive advantage. Quinn (1999) suggested that there are "core activities" that one firm will perform better than any other firm. In general, a function that is more core to an organization is less likely to be globally outsourced.

- Critical knowledge: Some data or knowledge must be under the control of the firm. In general, if a function provides critical knowledge, it is less likely to be globally outsourced.

- Impact on quality: The quality of the firm's services establishes its reputation and can create demand. If a firm is currently recognised in the industry for providing a high level of quality in a particular area, then global outsourcing in that area can harm quality. Quality is a relevant factor and can have either a positive or a negative influence on global outsourcing decisions. (Anderson, 1997).

- Flexibility: Flexibility includes demand flexibility, process flexibility and resource flexibility. Antonucci et al. (1998) noted that long contracts outsourced into a limited market have sometimes decreased flexibility. However, large enterprises may improve their flexibility via global outsourcing. In the literature, global outsourcing is used as a strategic driver to increase flexibility.

- Cost: In the literature, cost is the main reason for global outsourcing decisions. If the firm prefers to outsource a function for cost reasons, then it can be assumed that the current expenditures associated with that function are higher than the expected cost of purchasing the service. However, whether savings will actually accrue from global outsourcing is extremely uncertain. Sometimes, the reported cost savings may not be as high as was expected.

- Characteristics of the functions outsourced or kept in-house: In general, the more complex a function the less of a candidate it is for global outsourcing.

- Integration: Integration refers to the degree to which function is linked to other functions and systems within the enterprise. The more integrated the function, the more interactions and communication channels there are to maintain and monitor. Therefore, a function that is highly integrated is less of a candidate for global outsourcing.

Firms establish and execute global outsourcing plans in an effort to match competitors' attempts at outsourcing; improve non-competitive cost structures; focus on core competencies; reduce capital investment and overall fixed costs; achieve cost-competitive growth within their supply base for goods, services and technologies in the value chain; and establish future sales footprints in low-cost countries by outsourcing basic goods or business processes (Carter et al., 2008). An effective global sourcing strategy requires continual efforts to streamline manufacturing without sacrificing marketing flexibility (Kotabe & Murray, 2004).

According to the literature, firms prefer global outsourcing for the following reasons:

- Strategic focus / reduction of assets: Through global outsourcing activities, an enterprise can reduce its level of asset investment in manufacturing and related areas. Furthermore, global outsourcing can help management teams to redirect their attention to core competencies rather than focusing on maintaining a wide range of competencies (Kotabe et al., 2008).

- Supplementary power / lower production costs: Global suppliers are highly specialized in their own business, which lowers both their production costs for those of the firms that are outsourcing their business to them. Therefore, global outsourcing can decrease overall costs if firms globally outsource non-core activities (Quinn, 1999)

- Strategic flexibility: Global outsourcing can enhance a firm's strategic flexibility (Harris et al., 1998). If a firm is faced with a crisis in an external environment, it can simply change the volume of globally outsourced products it purchases. If the same product is outsourced to another firm within the home country, the firm will need to pay high reconstructing costs and may not respond quickly to the external environment.

- Relationship: Certain relationships with global suppliers can deliver competitive advantage for firms (Kotabe et al., 2008). Misunderstandings between buyer and suppliers may decrease a firm's level of performance (Carter et al., 2008).

Kotabe et al. (2008) suggested that an inverted-U shaped relationship exists between profitability and the degree of outsourcing. On the inverted-U shaped curve, there is an optimal degree of outsourcing for a firm. If the firm moves' away from this optimal point, profitability decreases dramatically. In global outsourcing strategy, there are also some disadvantages of increasing total product cost. Unfortunately, through global outsourcing, the cost of transportation, communication and information-sharing may increase. Domestic purchasing strategies require only short lead times because they reduce communication and transportation time requirements. The literature suggests that this may be the key reason why some enterprises do not prefer global outsourcing (Dana et al., 2007). The literature suggests some disadvantages of global outsourcing. These disadvantages can be seen below:

- The scope of the functions: If there are important interfaces between activities, decoupling them into separate activities performed by separate suppliers will generate less than optimal results and potential integration problems (Kotabe et al., 2008).

- Competition loss: Firms that engage in excessive outsourcing are essentially hollowing out their competitive base (Kotabe, 1998). Furthermore, an enterprise may lose negotiating power with its suppliers because the capabilities of the latter will increase relative to those of the former (Kotabe et al., 2008).

- Opportunistic behaviour: Global suppliers may behave opportunistically. Opportunistic behaviour allows a supplier to extract more rents from the relationship than it would normally do, for example, by supplying products of a lower quality than was previously agreed upon or by withholding information regarding changes in production costs (Kotabe et al., 2008).

- Limited learning and innovation: Suppliers may capture the critical knowledge by performing the activity. This situation is always a problem between buyer and supplier because both try to obtain all the individual benefits. Appropriation of innovations and rents is always a problem in such a complex buyer–supplier relationships (Nooteboom, 1999)

- Negative impact of exchange rates: Higher procurement costs can be seen by the negative impact of fluctuating exchange rates. During the Asian financial crisis, many foreign firms operating in Asian countries learned an invaluable lesson on the negative impact of fluctuating currency exchange rates on their procurement costs and profitability (Kotabe et al., 2008).

According to Kremic et al. (2006), the global outsourcing literature has referenced the following risks of global outsourcing: the potential for both unrealized savings and increased costs, employee morale problems, overdependence on suppliers, lost corporate knowledge and future opportunities, and under-satisfied customers. Additionally, global outsourcing may fail because the requirements of the relationship are inadequately defined because of a poor contract, a lack of guidance regarding planning or managing outsourcing initiatives, or poor supplier relations. Dana et al. (2007) cited lower production costs as the key advantage of a global outsourcing strategy, with poor control of quality being the main disadvantage. Lowe et al. (2002) addressed two risks of global outsourcing: fluctuations in exchange rates and relative rates of inflation in different countries. The impact of fluctuations in exchange rates can be analyzed in different ways, and these disparate analyses can yield different results. Brush et al. (1999) stated that many enterprises do not discuss exchange rates as a key factor in global outsourcing. Kouvelis (1999) stated that because of the high cost of switching global suppliers, purchasing managers do not switch suppliers until the effect of exchange rate fluctuation is extremely high. Vidal & Goetschalckx (2000) indicated that the impact of exchange rate fluctuation on overall cost is high. Under competitive pressure, many U.S. multinational companies globally outsource components and finished products to countries such as China, South Korea, Taiwan, Singapore, Hong Kong, and Mexico. Those countries are also known as low-cost countries (Kotabe & Murray, 2004). Firms in the US and the EU make different choices when selecting global outsourcing locations. In the US, 23% of enterprises prefer China, 14% prefer India, 10% prefer Mexico, 9% prefer Argentina and 8% prefer Brazil. In the EU, 19% prefer China, 14% prefer the Czech Republic, 12% prefer Poland and 10% prefer Hungary (Timmermans, 2005). The preferences of US and EU firms indicate what is known as "lowcost country sourcing" in the literature. Low-cost country sourcing entails the sourcing of services or functions from low-cost countries with lower labour and material costs. In recent years, low-cost country sourcing has created opportunities for purchasing managers (Carter et al., 2008). Sourcing from global suppliers can be risky, especially when the projected quality of the outsourced products is unknown. Motwani et al. (1999) noted that as the low-cost countries develop, the quality of the products produced in those countries will likely increase. As a result, firms that choose to forge relationships in these low-cost countries now through sourcing and purchasing may have an edge in these markets in the future. Although they may encounter challenges at first, the advantages that they enjoy in the future could outweigh these problems. This may be especially true for firms that aim to be truly global. Although the main factor driving global outsourcing is lower costs, experienced purchasing managers consider many factors simultaneously in making the decision to outsource internationally. According to the relevant literature, lower labour cost is not the key factor for many US enterprises that engage in global/domestic outsourcing (Sarkis & Talluri, 2002).

OUTSOURCING METHODS IN LITERATURE

Outsourcing has been widely discussed in the literature. There are several papers on supplier selection and global outsourcing in information technology (IT) and for service systems. To the best of our knowledge, few studies have been conducted on global outsourcing in manufacturing or production systems. This section is divided into two subsections: one that addresses general supplier selection methods in the literature and another that addresses global outsourcing methods in the literature.

General Supplier selection Methods in the Literature

There are several methods of general supplier selection presented in the literature. Categorical methods are qualitative models. Based on the buyer's experience and historical data, suppliers are evaluated using a particular set of criteria. The evaluations involve categorizing the supplier's performance as 'positive', 'neutral' or 'negative' with reference to a series of criteria (Boer et al., 2001). After a supplier has been rated for all the criteria, the buyer provides an overall rating, allowing the suppliers to be sorted into three categories. Data Envelopment Analysis (DEA) is concerned with the efficiency of decision making. The DEA method helps buyers to classify suppliers into two categories: efficient suppliers and inefficient suppliers. Liu et al. (2000) used DEA in the supplier selection process. They evaluated the overall performance of suppliers using DEA. Saen (2007) used IDEA (Imprecise Data Envelopment Analysis) to select the best suppliers based on both cardinal and ordinal data. Wu et al. (2009) proposed an augmented DEA approach to supplier selection. Songhori et al. (2011) presented a structured framework for helping decision makers to select the best suppliers for their firm using DEA. Cluster Analysis (CA) is a class of statistical techniques that can be used with data that exhibit "natural" groupings (Boer et al., 2001).

Case-Based Reasoning systems (CBR) combine a cognitive model describing how people use and reason from past experience with a technology for finding and presenting experience (Choy et al., 2003-a). Choy et al. (2002-b) enhanced a CBR-based supplier selection tool by combining the Supplier Management Network (SMN) and the Supplier Selection Workflow (SSW). Choy et al. (2005) used CBR to select suppliers in a new product development process. In linear weighting, the criteria are weighted, and the criterion with the largest weight has the greatest importance. The score for a particular supplier is based on the criteria and their different levels of importance, and some criteria have a high degree of precision. Ghodsypour & O'Brien (1998) integrated the Analytic Hierarchy Process (AHP) and linear programming to consider both tangible and intangible factors in choosing the best suppliers and the optimum order quantities. Lee et al. (2001) used only the AHP for supplier selection. They determined the supplier selection criteria based on purchasing strategy and criterion weights using the AHP. Liu & Hai (2005) used DEA to determine the supplier selection criteria. They then interviewed 60 administrators to determine the priority level of the criteria and used the AHP to select suppliers. Ting & Cho (2008) presented a two-step decision-making procedure. They used the AHP to select a set of candidate suppliers for a firm and then used a Multi-Objective Linear Programming (MOLP) model to determine the optimal allocation of order quantities to those suppliers. Boer et al. (1998) used the ELECTRE 1 technique to evaluate the five supplier candidates. Xia & Wu (2007) used an integrated approach to the AHP, which was improved using rough set theory and multi-objective mixed integer programming to simultaneously determine the number of suppliers to employ and the order quantities to be allocated to these suppliers in the case of multiple sourcing and multiple products. Multiple criteria and supplier capacity constraints were both taken into account. Wang et al. (2004) used an integrated AHP and preemptive goal programming (PGP)-based multi-criteria decisionmaking process to analyze both the qualitative and quantitative factors guiding supplier selection. Liu and Hai (2005) compared the use of the Voting Analytic Hierarchy Process (VAHP) and the use of the AHP for supplier selection. Chan & Kumar (2007) identified some of the important decision criteria, including risk factors in developing an efficient system of global supplier selection. They used the Fuzzy Extended Analytic Hierarchy Process (FEAHP) to select suppliers. Chan & Chan (2010) used an AHP-based model to solve the supplier evaluation and selection problem for the fashion industry. Kumar & Roy (2011) proposed the use of a rule-based model with the AHP to aid decision makers in supplier evaluation and selection. Total Cost of Ownership models (TCO) include all costs related to the supplier selection process that are incurred during a purchased item's life cycle. Degraeve & Roodhooft (1999) evaluated suppliers based on quality, price and delivery performance using TCO. They emphasised that uncertainty related to demand, delivery, quality and price must be reflected in the decision problem. Ramanathan (2007) proposed the integrated DEA-TCOAHP model for supplier selection. According to Boer et al. (2001), Mathematical Programming models (MP) allows the decision maker to formulate the decision problem in terms of a mathematical objective function that must subsequently be maximized and minimized by varying the values of the variables in the objective function. MP models are more objective than rating models because they force the decision maker to explicitly state the objective function, but MP models often only consider more quantitative criteria. Karpak et al. (1999) developed a supplier selection tool that minimizing costs and maximizing quality reliability. Ghodsypour & O'Brien (1998) integrated the AHP and

Linear Programming (LP) models. Their model presented a systematic approach that took into account both qualitative and quantitative criteria. They also developed sensitivity algorithms for different scenarios. Ghodsypour & O'Brien (2001) used mixed integer programming, taking into account the total cost of logistics. Degraeve & Roodhooft (2000) computed the purchasing cost associated with different purchasing strategies using MP. Barla (2003) reduced the number of suppliers from 58 to 10 using the multi-criteria selection method. Hong et al. (2005) decomposed the supplier selection process into two steps. They used cluster analysis to preselect suppliers and then used MP to select the most appropriate supplier. Yang et al. (2007) studied a supplier selection problem in which a buyer facing random demand must decide the quantity of products it will order from a set of suppliers with different yields and prices. They provided the mathematical formulation for the buyer's profit maximization problem and proposed a solution method based on combining the active set method and the Newton search procedure. Kheljani et al. (2007) considered the issue of coordination between one buyer and multiple potential suppliers in the supplier selection process. In contrast, in the objective function in the model, the total cost of the supply chain is minimized in addition to the buyer's cost. The total cost of the supply chain includes both types of costs. The model was solved using mixed-integer nonlinear programming. Liao & Rittscher (2007) developed a multi-objective programming model, integrating supplier selection to procure lot sizing and carrier selection decisions for a single purchasing item over multiple planning periods during which the demand quantities are known but inconstant. Rajan et al. (2010) proposed a supplier selection model for use in a multiproduct, multi-vendor environment based on an integer linear programming model.

Artificial intelligence (AI)-based systems are computer-aided systems that can be trained using data on purchasing experience or historical data. The available types of AI-based supplier selection applications include Neural Networks (NN) and Expert Systems (ES). One of the important advantages of the NN method is that the method does not require the formulation of the decision-making process. As a result, NNs can cope better with complexity and uncertainty than traditional methods can; these systems are designed to be more similar to human judgment in their functioning. The system user must provide the NN with the properties of the current case. The NN provides information to the user based on what it has learned from the historical data. Albino & Garavelli (1998) further developed the neural network-based decision support system for subcontractor ratings in construction firms. The system includes a back-propagation algorithm. The constructed network is trained using examples so that the system does not require decision-making rules. Vokurka et al. (1996) and Wei et al. (1997) developed an expert system for supporting the supplier selection process. Chen et al. (2006) used linguistic values to assess the ratings and weights of various supplier selection factors. These linguistic ratings were expressed using trapezoidal or triangular fuzzy numbers. Then, they proposed the use of a hierarchy Multiple Criteria Decision-Making (MCDM) model based on fuzzy-sets theory to address supplier selection problems in the supply chain system. Wang & Che (2007) presented an integrated assessment model for manufacturers to use to solve complex product configuration change problems efficiently and effectively. The model made it possible to determine what fundamental supplier combination would best minimize the cost–quality score if and when proposed by the customer and/or engineer. The researchers combined fuzzy theory, T transformation technology, and genetic algorithms. Liao & Rittscher (2007) studied the supplier selection problem under stochastic demand conditions. Stochastic supplier selection is determined by simultaneously considering the total cost, the quality rejection rate, the late delivery rate and the flexibility rate while also taking into account constraints on demand satisfaction and capacity. The researchers used GA to solve the problem. Wang (2008) developed a decision-making procedure that could be used for supplier selection when product part modifications were necessary. The aim of the research was to determine acceptable near-optimal solutions within a short period of time using a solution-finding model based on Genetic Algorithms (GA). Aksoy & Öztürk (2011) presented a neural network-based supplier selection and supplier performance evaluation system for use in a just-in-time (JIT) production environment. Chang et al. (2011) proposed the use of a fuzzy decision-making method to identify evaluation factors that could be used for supplier selection. Jiang & Chan (2011) proposed a method of using a fuzzy set theory with twenty criteria to evaluate and select suppliers.

Global Outsourcing Methods in the Literature

Canel & Khumawala (1996) proposed a 0-1 mixed integer programming formulation model for international facilities location problems. They determined the location of the international facility and the capacity of that facility. The objective of the model is to maximize the after-tax profit. The proposed model includes different costs, including investment cost, fixed costs, transportation costs, shortage costs and holding costs. The researchers developed two different mathematical models: one for a capacitated case and the other for an uncapacitated case. They used demand and price as the deterministic parameters. Their research could be extended by relaxing the assumptions of deterministic demand, prices, costs, etc. within the problem and treating those factors as stochastic parameters. Huchzermeier & Cohen (1996) developed a stochastic dynamic programming formulation for evaluating global manufacturing strategy options while taking switching costs into account in a stochastic exchange-rate environment. The objective of the model is to maximize after-tax profits. The model includes taxes, fixed and variable costs, capacity and exchange rates. The decision variable in the model is production quantities. The researchers developed different scenarios for different exchange rates. Each model has its own solution. However, the model does not include qualitative parameters. Canel & Khumawala (1997) presented an efficient branch-and-bound procedure for solving uncapacitated, multi-period international facilities location problems. The branch-andbound problems can be solved using LINDO. The parameters of the model are assumed to be deterministic. Dasu & Torre (1997) presented a model for planning a global supply network for a multinational yarn manufacturer. The objective of the model is to maximize the overall profits of the global supply chain network. The model includes tariffs, exchange rates and transportation costs. The proposed model is non-linear, but the authors make some assumptions in the model to give it a linear structure. Kouvelis & Gutierrez (1997) solved the newsvendor problem in the textile industry for "style goods". The proposed model determines the production quantities while minimizing shortages and holding costs for a multiple-location manufacturer in a multiple-location market. Shortage costs in this context include the costs associated with lost sales, and holding costs includes the cost associated with excess inventory left over after the selling season. The model includes transportation costs, exchange rate uncertainty and stochastic demand uncertainty, but the model does not include global cost factors such as taxes and tariffs. The study evaluates alternative plans for supply chain design and centralized and decentralized production decision-making mechanisms. The researchers also stated that centralized production decision-making is superior to decentralized production decision-making but that application and control problems are associated with centralized coordination. The proposed model can be easily implemented by purchasing managers. The researchers noted that the production decision-making process can be affected by the uncertainty of global markets and that models with stochastic parameters (such as models for analysing political risk and exchange rate fluctuations) can be used in future research. Munson & Rosenblatt (1997) described local content rules and developed models for selecting global suppliers while satisfying local content provisions. The parameters of the model are deterministic, and the penalty for breaking local content rules is very high. The researchers used the mixed integer programming method to solve the model. The decision variables for the model are the selection of the global supplier and the allocation of orders among the selected suppliers. The objective of the model is to minimize purchasing, production, transportation and fixed costs. The model considers only costs and local content rules. The model does not take into account quantitative parameters or exchange rates.

Coman & Ronen (2000), formulated the global outsourcing problem as a linear programming (LP) problem, identified an analytical solution, and compared that solution with the solutions obtained using the standard cost accounting model and the theory of constraints. The decision variable for the model is the production quantity in terms of preference to manufacture versus preference to outsource. The solution attained indicated that linear programming yielded better results than the two other methods (standard cost accounting and the theory of constraints). Canel & Khumawala (2001) solved an international facilities location problem using the heuristic method. They developed 12 heuristic methods, but their models do not include quantitative parameters or exchange rates. Vidal & Goetschalckx (2001) presented a model for optimizing global supply that maximizes after-tax profits for a multinational corporation. The model includes transfer prices and the allocation of transportation costs as explicit decision variables. Transfer prices and flows between multinational facilities are calculated in the model. The model does not address the supplier selection problem. The model entails a nonconvex optimization problem with a linear objective function, a set of linear constraints, and a set of bilinear constraints. Because the resulting problem is NP hard, the researchers developed a heuristic successive linear programming solution procedure. Canel & Das (2002) proposed the use of a 0-1 mixed integer programming model to determine for particular time periods which countries a firm should choose as the locations of its global manufacturing facilities. The model was also used to determine the quantity to be produced at each global manufacturing facility and the quantities to be shipped from the global facilities to customers. The study had two goals: to determine the location of the global manufacturing facility and to develop a mathematical model that included global marketing and manufacturing factors. The proposed model does not include quantitative parameters or take exchange rates into account. Hadjinicola & Kumar (2002) presented a model that includes manufacturing factors together with factory location, inventory, economies of scale, product design, and postponement. In the study, different manufacturing and marketing strategies were evaluated for two different countries. The proposed model is a descriptive model; therefore, the decision variables are not entirely clear. The model can be used to evaluate different manufacturing and marketing strategies in terms of cost and profit functions. The model also includes the effect of exchange rates. Lowe et al. (2002), proposed a model that included exchange rates, using it to choose the location and capacity of global manufacturing facilities in the chemical industry. They analysed data from the year 1982, setting the production capacity, purchasing volume, capacity, fixed and variable costs, transportation costs and taxes for each production facility. To take into account the effect of exchange rates, they reviewed historical data from 22 years and constructed nine different scenarios, calculating the costs for each scenario. They developed a two-stage method of solving the problem. The first stage is a short planning period (as an example 1 year), and the second stage is an optional stage that can be used for long-term planning. However, for longer periods, the first stage can also be repeatedly used. Teng & Jaramillo (2005) presented a model for global supplier selection in the textile industry. They weighted their criteria and sub-criteria based on expert opinions. The overall scores for the global supplier are calculated by multiplying the score and the weight of the criteria. Goh et al. (2007) presented a stochastic model of the multi-stage global supply chain network problem, incorporating a set of related risks: supply, demand, exchange, and disruption. The objective of the model is to maximize after-tax profits. The model includes demand uncertainty, exchange rates, taxes and tariffs. The researchers composed different scenarios for demand uncertainty and exchange rates and presented different clusters of stochastic parameters. Because the proposed model is a convex linear model, the authors relaxed certain parameters to make the problem linear, but theirs is a descriptive model that is not applied to a real-world scenario. Lin et al. (2007) proposed the use of a decision model to support global decision making. The model uses two multiple-criteria decision aid techniques (the AHP and PROMETHEE II) and incorporates multiple dimensions (infrastructure, country risk, government policy, value of human capital and cost) into a sensitivity analysis. The authors used the AHP to determine the weight of the criteria and used PROMETHEE to select global suppliers based on weighted criteria. Kumar & Arbi (2008) used a simulation model to forecast lead time and total cost in a global supply environment. Based on their results, it seems that important cost savings can be attained through global outsourcing. However, lead time is an essential factor in real life. The authors stated that global outsourcing is not a viable way to meet short-term market demands but that for large seasonal orders, global outsourcing can be a significant cost-saver. Ray et al. (2008) described the cause of the outsourcing problem, formulated it as a linear programming problem, developed a corresponding function, and offered a simplified criterion for ordering products in terms of preference for manufacturing versus preference for global outsourcing. The authors used a hybrid approach, incorporating the Hurwicz criterion, the theory of constraints (TOC) and linear programming. Some weaknesses of the proposed method are that it is difficult to change the traditional cost accounting system, that it would take time to implement the approach, and that people may be reluctant to use the approach because it requires them to justify their preferences rather than simply saying yes or no. It also requires a new decision-making process.

Wang et al. (2008a) divided firm activities into core activities, core-close activities, coredistinct activities and disposable activities. They used the ELECTRE 1 method to determine which of those four activities can be globally outsourced. Wang et al. (2008b) presented a model for the global outsourcing of logistics activities. They determined the evaluation criteria, ranked the criteria using the AHP and constructed a method using PROMETHEE for global supplier selection. Feng & Wu (2009) presented several tax-saving approaches and developed a tax savings model for maximizing after-tax profit from logistics activities by global manufacturers. Using this model, logistics activities are evaluated in terms of tax savings. It has been observed that the tax saving model has dramatically increased manufacturer profits. The suppliers are selected based on tax savings, whereas any other criteria are disregarded. Ren et al. (2009) treated global supply chains as agile supply chains and explained agility as the ability to change and adapt quickly to changing circumstances. The model facilitates supplier selection for agile supply chains. The authors determined 10 criteria and 32 subcriteria for supplier selection. The weights of the criteria were determined based on expert opinions and used to rank the suppliers. Perron et al. (2010) presented a mathematical model for multinational enterprises to use to determine transfer prices and the flow of goods between global facilities. The model includes bilinear constraints; therefore, the authors relaxed the constraints to simplify the model. They developed a branch-and-cut algorithm and two different heuristics to solve the model. The heuristic methods can be summarised as follows:

- Variable Neighbour Search Method (VNS): This method is based on the concept of systematic changes in neighbourhoods during the search. VNS explores nearby and then increasingly far neighbourhoods for the best-known solution in a probabilistic fashion. Therefore, often favourable characteristics of the best-known solution will be kept and used to obtain promising neighbouring solutions. VNS was repeatedly used with a local search routine to transition from these neighbouring solutions to local optima.

- Alternate Heuristic (ALT): Given two subsets of variables, the ALT heuristic solves the problem by alternately fixing the variables of one of the subsets. The subsets of variables must be such that the model becomes linear when fixing the variables of one of the subsets. When one of these linear programs is solved, its solution becomes a set of parameters in the other one. ALT can be converging to local optima.

The objective of the model is to maximize after-tax profit given taxes, capacity, transfer prices and demand. Satisfactory results were reported when small problems were solved using heuristic methods.

OUTSOURCING DECISION CRITERIA

To be competitive in the global market, firms must attain the knowledge necessary to systematically evaluate all potential suppliers and select the most suitable ones. The factors most often used in current supplier evaluation are quality, supplier certification, facilities, continuous improvement, physical distribution and channel relationships (Weber, 1991).

In the supplier selection process, it is not always easy to recognize precise rules, but there is, in general, a coherent way to solve the problem. This coherence can be rooted in intuition, experience, common sense, or inexplicable rules. Supplier rating is then a problem usually solved by subjective criteria, based on personal experiences and beliefs, on the available information and, sometimes, on techniques and algorithms supporting the decision process (Albino & Garavelli, 1998). The key to enhancing the quality of decision making in supplier selection include the powerful computer-related concepts, tools and techniques that have become available in recent years (Wei et al., 1997). Chao et al. (1993) concluded that quality and on-time delivery are the most important attributes of purchasing performance. Ghodsypour & O'Brien (1998) agreed that cost, quality and service are the three main factors that should influence supplier selection. Brigs (1994) stated that joint development, culture, forward engineering, trust, supply chain management, quality and communication are the key requirements of supplier partnerships apart from optimum cost. Petroni & Braglia (2000) evaluated the relative performance of suppliers with multiple outputs and inputs, considering management, production facilities, technology, price, quality, and delivery compliance. Wei et al. (1997) examined factors such as supply history, product price, technological ability and transport cost. Making sourcing decisions based on delivery speed and cost is the best way to improve performance (Tan 2001). Global outsourcing reduces the fixed investment costs of a firm in its own economic region. Today, in making global outsourcing decisions, many enterprises also consider quality, reliability, and technology when evaluating the components and products to be procured (Kotabe & Murray, 2004) rather than only considering price. In developing global outsourcing strategies, firms must consider not only manufacturing costs, the costs of various resources, and exchange rate fluctuations but also the availability of infrastructure (including transportation, communications, and energy), industrial and cultural environments, and ease of working with foreign host governments, among others (Kotabe et al., 2008). Several factors influence global outsourcing decisions. Canel & Das (2002) outlined the factors that most commonly influence global manufacturing facility locations. Those factors are labour and other production inputs, political stability, the attitude of the host government towards foreign investment, host government tax and trade policies, proximity to major markets, access to transportation and the existence of other competitors. Choy et al. (2005) stated that good customer–supplier relationships are necessary for an organization to respond to dynamic and unpredictable changes. They considered price, delivery, quality, innovation, technology level, culture, commercial awareness, production flexibility, ease of communication and current reputation when selecting and evaluating suppliers. Teng & Jaramillo (2005) used five main criteria and 20 sub-criteria for global supplier selection in the textile-apparel industry. Those criteria are delivery (geographic location, freight terms, trade restrictions and total order lead time), flexibility (capacity, inventory availability, information sharing, negotiability and customization), cost (supplier selling price, internal costs, ordering and invoicing), quality (continuous improvement programs, certification, customer service and the percentage of on-time shipments), reliability (feelings of trust, the national political situation, the status of the currency exchange and warranty policies).

Narasimhan et al. (2006) composed a model, for global supplier selection and order allocation and considered criteria such as direct product cost, the indirect cost of coordination, quality, delivery reliability and complexity of the supply base. Lin et al. (2007) used infrastructure, country risk, government policy, human capital and cost when evaluating global suppliers. Carter et al. (2008) analysed the low-cost countries and their capabilities. They evaluated the factors such as labour cost, work ethic, intellectual property, market attraction, delivery reliability, reliable transportation, transportation costs, government support for business, political stability, flexibility, predictable border crossing and corruption in 12 different low-cost countries using perceptual mapping. They stated that experienced purchasing managers not only consider cost but also conduct a multicriteria evaluation of global outsourcing decisions. Au & Wong (2008) identified four main categories of factors from the literature: cost (labour costs, material costs and transportation costs), product quality (technological capabilities, reliability and trust), time to market (geographical proximity and transportation time) and country factors, including both internal country factors (such as infrastructure and ethical issues) and external country factors (such as the political and economical situation and social, linguistic and cultural differences). Chan et al. (2008) examined the decision variables influencing global supplier selection and identified five main criteria and 19 sub-criteria: total cost of ownership (product cost, total logistics management cost, tariffs and taxes), product quality (conformance with specifications, product reliability, quality assessment techniques and process capabilities), service performance (delivery reliability, information sharing, flexibility, responsiveness and customer responses), supplier background (technological capabilities, financial status, facilities, infrastructure and market reputation), and risk factors (geographical location, political stability and foreign policies, exchange rates and economic position, terrorism and the crime rate). Ku et al. (2010) identified the following criteria as important to global supplier selection: cost (product price, freight costs and custom duties), quality (rejection rate, process capabilities and quality assessments), service (on-time delivery, technological support, responses to changes and ease of communication), risk (geographical location, political stability and the status of the economy). Ku et al. (2010) suggested that qualitative criteria (e.g., the characteristics of the purchased items) be considered in future research.

CONCLUSION

Manufacturers must be the forerunners to be competitive in today's global markets. That is why manufacturers must keep in touch with the dynamic requirements of the market and be receptive to reforms. An increasing proportion of raw materials and work-in-process (WIP) for manufactured products is sourced globally by multinational manufacturers in today's industries. To become a world-class manufacturer, a firm must not only compete globally in the marketplace but also be competitive and consistent in terms of costs, technological leadership, and quality. High-quality inputs are becoming the focus of many purchasing departments.

The design of global supply chains has been a challenging optimization problem for many years. In a continuing effort to remain competitive, many firms are considering new sources for their raw materials and components, new locations for their production and distribution facilities, and new markets in which to sell their products without regard for national boundaries. The design of global supply chains has been a challenging optimization problem for many years. The current globalization of the economy is forcing firms to design and manage their supply chains efficiently on a worldwide basis. It is well known that large enterprises no longer operate in a single market. In seeking to penetrate global markets and obtain their benefits, firms are under excess pressure to reduce the price of their products and thus their production and material costs. According to the literature, global outsourcing is mainly analyzed in terms of cost. During the evaluation of global outsourcing, neither multi-criteria evaluations nor qualitative assessments are always made, but many decision criteria must be considered when configuring a global supply chain system. Historically, labour cost has been one of the most decisive factors in global outsourcing decision making. Recently, the rapidly changing business environment has increased pressure on decision makers to properly analyze the relevant decision criteria. Strategic decision making requires the use of tangible, intangible, strategic and operational decision criteria. There are many factors that need to be considered simultaneously, and purchasing managers need a structured method of using these criteria in their decision making. The AHP is the widely used a multi-criteria, decision-making method used by academicians and practitioners for supplier selection and global outsourcing decision making. The AHP is based on dual comparisons of decision criteria. As the number of decision criteria increases, the complexity of the system also increases. Mathematical models are not sufficient for evaluating qualitative criteria. One of the major factors complicating the modelling of global outsourcing decision making is "uncertainty". Exchange rate fluctuations, variable transportation times, demand uncertainty, the variability of market prices, and political instability are among the most important sources of uncertainty. An effective decision-making methodology for global outsourcing must address those uncertainties. In recent years, artificial intelligence tools such as neural networks, fuzzy logic and genetic algorithms have been increasingly used for outsourcing decision making. Those methods are more appropriate under uncertainty and can better address qualitative criteria.

REFERENCES

1. Aksoy, A., & Öztürk, N. (2011). Supplier Selection And Performance Evaluation In Just-InTime Production Environments. Expert Systems with Applications, Vol. 38, No. 5, pp. 6351-6359, ISSN: 0957-4174

2. Albino, V., & Garavelli, A.C. (1998). A Neural Network Application to Subcontractor Rating in Construction Firms. International Journal of Project Management, Vol. 16, No. 1, pp.9-14, ISSN: 0263-7863

3. Anderson, M.C., (1997). A Primer in Measuring Outsourcing Results. National Productivity Review, Vol. 17, No.1, pp. 33-41, ISSN 0277-8556

4. Antonucci, Y.L., Lordi, F.C., & Tucker J.J. (1998). The Pros and Cons of IT Outsourcing. Journal of Accountancy, Vol.185, No.6, ISSN: 1945-0729

5. Au, K.F., & Wong, M. C. (2008). Decision Factors In Global Textile And Apparel Sourcing After Quota Elimination. The Business Review, Vol.9, No.2, pp. 153-157, ISSN 1540 -1200

6. Barla, S.B. (2003). A Case Study Of Supplier Selection For Lean Supply By Using A Mathematical Model. Logistics Information Management, Vol.16, No.6, pp. 451-459,ISSN 0957-6053

7. Boer, L., Labro, E., & Morlacchi, P. (2001). A Review of Methods Supporting Supplier Selection. European Journal of Purchasing & Supply Management, Vol.7, pp. 75-89,ISSN: 1478-4092.

8. Boer, L., Wegen, L., & Telgen, J. (1998). Outranking Methods In Support

Of Supplier Selection. European Journal of Purchasing & Supply Management, Vol.4, pp. 109-118,ISSN: 1478-4092

9. Briggs, P. (1994). Case Study: Vendor Assessment For Partners In Supply. European Journal of Purchasing and Supply Management, Vol. 1, No. 1, pp. 49–59, ISSN 0969-7012

10. Brush, T.H., Martin, C.A., & Karnani, A. (1999). The Plant Location Decision in Multinational Manufacturing Firms: An Empirical Analysis of InternationalBusiness and Manufacturing Strategy Perspectives. Production and Operations Management, Vol.8, pp.109-132, ISSN: 1937-5956

11. Canel, C., & Das, S.R. (2002). Modeling Global Facility Location Decisions: Integrating Marketing And Manufacturing Decisions. Industrial Management and Data Systems, Vol.102, No.2, pp. 110-118, ISSN 0263-5577

12. Canel, C., & Khumawala, B.M. (2001). International Facilities Location: A Heuristic Procedure For The Dynamic Uncapacitated Problem. International Journal of Production Research, Vol.39, No.17, pp. 3975-4000, ISSN 0020-7543

13. Canel, C., & Khumawala, B.M. (1997). Multi-Period International Facilities Location: An Algorithm And Application. International Journal of Production Research, Vol.35,No.7, pp. 1891- 1910, ISSN 0020-7543

14. Canel, C. & Khumawala, B.M. (1996). A Mixed-Integer Programming Approach For The International Facilities Location Problem. International Journal of Operations & Production Management, Vol.16, No.4, pp. 49-68, ISSN 0144-3577

15. Carter, J.R., Maltz, A., Yan, T., & Maltz, E. (2008). How Procurement Managers View Low Cost Countries and Geographies. International Journal of Physical Distribution &Logistics Management, Vol.38, No.3, pp. 224-243, ISSN: 0960-0035

16. Chan F.T.S., & Chan H.K. (2010). An AHP Model For Selection Of Suppliers In The FastChanging Fashion Market. International Journal of Advanced Manufacturing Technology, Vol. 51, pp. 1195–1207, ISSN 0268-3768

17. Chan, F.T.S., & Kumar, N. (2007). Global Supplier Development Considering Risk Factors Using Fuzzy Extended AHP-Based Approach. The International Journal of Management Science, Vol.35, pp. 417-431, ISSN 0305-0483

18. Chan, F.T.S., Kumar, N., Tiwari, M. K., Lau, H. C. W., & Choy, K. L. (2008). Global Supplier Selection: A Fuzzy-AHP Approach. International

Journal of Production Research,Vol.46, No.14, pp. 3825–3857, ISSN: 0020-7543

19. Chang B., Chang C., & Wu C. (2011). Fuzzy DEMATEL Method For Developing Supplier Selection Criteria. Expert Systems with Applications, Vol. 38,pp. 1850–1858, ISSN 0957-4174

20. Chao, C., Scheuing, E.E., & Ruch, W.A. (1993). Purchasing performance evaluation: an investigation of different perspectives. International Journal of Purchasing and Materials Management, Vol.29, No.3, pp. 33–39, ISSN 1055-6001

21. Chen, C.T., Lin, C.T., & Huang, S.F. (2006). A Fuzzy Approach For Supplier Evaluation And Selection In Supply Chain Management. International Journal of Production Economics, Vol.102, pp.289–301, ISSN: 0925-5273

22. Choy, K.L., Lee, W.B., Lau, H.C.W., & Choy, L.C. (2005). A Knowledge Based Supplier Intelligence Retrieval System For Outsource Manufacturing. Knowledge-Based System, Vol.18, pp. 1-17, ISSN 0950-7051

23. Choy, K.L., Lee, W.B., & Lo, V. (2003). Design Of A Case Based Intelligent Supplier Relationship Management System- The Integration Of Supplier Rating System And Product Coding System. Expert Systems with Applications, Vol. 25, pp. 87-100, ISSN:0957-4174

24. Choy, K.L., Lee, W.B., & Lo, V. (2002). On The Development Of A Case Based Supplier Management Tool For Multi-National Manufacturers. Measuring Business Excellence, Vol.6, No.1,pp. 15-22, ISSN : 1368-3047

25. Coman, A., & Ronen, B. (2000). Production Outsourcing: A Linear Programming Model For The Theory-Of-Constraints. International Journal of Production Research, Vol. 38, No.7, pp. 1631-1639, ISSN: 0020-7543

26. Dana, L.P., Hamilton, R. T., & Pauwels, B. (2007). Evaluating Offshore and Domestic Production in the Apparel Industry: The Small Firm's Perspective. Journal of International Entrepreneurship, Vol.5, pp. 47–63, ISSN1570-7385

27. Dasu, S., & Torre, J. (1997). Optimizing an International Network of Partially Owned Plants Under Conditions of Trade Liberalization. Management Science, Vol.43, No.3, pp.313-333, ISSN: 0025-1909

28. Degraeve, Z., & Roodhooft, F. (2000). A Mathematical Programming Approach For Procurement Using Activity Based Costing. Journal of Business Finance and Accounting, Vol. 27, No.1&2, pp. 69-98, ISSN

1468-5957

29. Degraeve, Z., & Roodhooft, F. (1999). Improving The Efficiency Of The Purchasing Process Using Total Cost Of Ownership Information: The Case Of Heating Electrodes At Cockerill Sambre S.A.. European Journal of Operational Research, Vol.112, pp. 42-53,ISSN 0377-2217

30. Feng, C.M., & Wu, P.J. (2009). A Tax Savings Model For The Emerging Global Manufacturing Network. International Journal of Production Economics, Vol. 122, No.2, pp. 534-546, ISSN 0925-5273

31. Ghodsypour, S.H., & O›Brien, C. (1998). A Decision Support System For Supplier Selection Using An Integrated Analytic Hierarchy Process And Linear Programming. International Journal of Production Economics, Vol. 56-57, pp. 199-212, ISSN 0925-5273

32. Ghodsypour, S.H., & O›Brien, C. (2001). The Total Cost Of Logistics In Supplier Selection, Under Conditions Of Multiple Sourcing, Multiple Criteria And Capacity Constraints. International Journal of Production Economics, Vol. 73, pp. 15-27, ISSN 0925-5273

33. Gillett, J. (1994). Viewpoint. The Cost-Benefit of Outsourcing: Assessing The True Cost of Your Outsourcing Strategy. European Journal of Purchasing & Supply Management, Vol. 1, No. 1, pp. 45-7, ISSN: 1478-4092

34. Goh, M., Lim, J.Y.S., & Meng, F. (2007). A Stochastic Model For Risk Management In Global Supply Chain Networks. European Journal of Operational Research, Vol.182, pp. 164–173, ISSN 0377-2217

35. Gottfredson, M., Puryear, R., & Phillips, S. (2005). Strategic Sourcing From Periphery to the Core. Harvard Business Review, February, ISSN: 0017-8012

36. Hadjinicola, G.C., & Kumar, K.R. (2002). Modeling Manufacturing And Marketing Options In International Operations. International Journal of Production Economics, Vol.75, pp.287–304, ISSN: 0925-5273

37. Harris, A., Giunipero, L.C. , Tomas, G., & Hult, M. (1998). Impact of Organizational and Contract Flexibility on Outsourcing Contracts. Industrial Marketing Management,Vol.27, pp. 373–384, ISSN: 0019-8501

38. Hong, G.H., Park, S.C., Jang, D.S., & Rho, H.M. (2005). An Effective Supplier Selection Method For Constructing A Competitive Supply-Relationship. Expert Systems with Applications, Vol. 28, pp. 629-639, ISSN: 0957-4174

39. Huchzermeier, A., & Cohen, M.A. (1996). Valuing Operational Flexibility

under Exchange Rate Risk. Operations Research, Special Issue on New Directions in Operations Management, Vol.44, No.1, pp 100-113, ISSN 0030-364X

40. Jiang W., & Chan F.T.S. (2011). A New Fuzzy Dempster MCDM Method And Its Application In Supplier Selection. Expert Systems with Applications, Vol. 38, No.8, pp. 9854-9861,ISSN 0957-4174

41. Karpak, B., Kasuganti, R.R., & Kumcu, E. (1999). Multi-Objective Decision-Making In Supplier Selection: An Application Of Visual Interactive Goal Programming. Journal of Applied Business Research, Vol.15, No.2, pp. 57-71, ISSN 0892-7626

42. Kheljani, J.G., Ghodsypour, S.H., & O'Brien, C. (2007). Optimizing Whole Supply Chain Benefit Versus Buyer's Benefit Through Supplier Selection. International Journal of Production Economics, Vol. 121, No.2, pp. 482-493, ISSN: 0925-5273

43. Kotabe, M. (1998). Efficiency vs. Effectiveness Orientation of Global Sourcing Strategy: A Comparison of U.S. and Japanese Multinational Companies. Academy of Management Executive, Vol. 12, No. 4, pp. 107-119, ISSN: 1079-5545

44. Kotabe, M., Mol, M.J., & Murray, J.Y. (2008). Outsourcing, Performance, and The Role of ECommerce:A Dynamic Perspective. Industrial Marketing Management, Vol.37, pp.37–45, ISSN: 0019-8501

45. Kotabe, M., & Murray, J.Y. (2004). Global Sourcing Strategy and Sustainable Competitive Advantage. Industrial Marketing Management, Vol. 33, pp. 7– 14, ISSN: 0019-8501

46. Kouvelis, P., & Gutierrez, G.J. (1997). The Newsvendor Problem in a Global Market: Optimal Centralized and Decentralized Control Policies For A Two Market Stochastic Inventory System. Management Science, Vol.43, No.5, pp.571-585, ISSN: 0025-1909

47. Kremic, T., Icmeli Tukel, O., & Rom, W. O. (2006). Outsourcing Decision Support: A Survey Of Benefits, Risks, And Decision Factors. Supply Chain Management: An International Journal, Vol. 11, No.6, pp. 467–482, ISSN: 1359-8546

48. Ku, C.Y., Chang, C.T., & Ho, H.P. (2010). Global Supplier Selection Using Fuzzy Analytic Hierarchy Process And Fuzzy Goal Programming. Quality & Quantity, Vol. 44, No.4, pp. 623-640, ISSN: 0033-5177

49. Kumar, S., & Arbi, A.S. (2008). Outsourcing Strategies For Apparel Manufacture: A Case Study. Journal of Manufacturing Technology Management, Vol.19, No.1, pp. 73-91, ISSN 1741-038x

50. Kumar J., & Roy N. (2011). Analytic Hierarchy Process (AHP) For A Power Transmission Industry To Vendor Selection Decisions. International Journal of Computer Applications, Vol.12, No.11, pp. 26-30, ISSN 0975 - 8887

51. Lee, E.K., Sungdo, H., & Kim, S.K. (2001). Supplier Selection And Management System Considering Relationships In Supply Chain Management. IEEE Transactions on Engineering Management, Vol.48, No.3, pp. 307-318, ISSN: 0018-9391

52. Levy, D.L. (2005). Offshoring in the New Global Political Economy. Journal of Management Studies, Vol.42, No.3, pp. 685-693, ISSN 1467-6486

53. Liao, Z., & Rittscher, J. (2007). Integration Of Supplier Selection, Procurement Lot Sizing And Carrier Selection Under Dynamic Demand Conditions. International Journal of Production Economics, Vol.107, pp. 502–510, ISSN: 0925-5273

54. Lin, Z.K., Wang, J.J., & Qin, Y.Y. (2007). A Decision Model For Selecting An Offshore Outsourcing Location: Using A Multicriteria Method. Proceedings of the 2007 IEEE International Conference on Service Operations and Logistics, and Informatics, pp. 1-5,

55. ISBN 978-1-4244-1118-4, Philadelphia, PA, USA, August 27-29, 2007 Liu, J., Ding, F.Y., & Lall, V. (2000). Using Data Envelopment Analysis To Compare Suppliers For Supplier Selection And Performance Improvement. Supply Chain Management: An International Journal, Vol. 5, No.3, 143-150, ISSN: 1359-8546

56. Liu, F.F.H., & Hai, H.L. (2005). The Voting Analytic Hierarchy Process Method For Selecting Supplier. International Journal of Production Economics, Vol.97, No.3, pp. 308-317, ISSN 0925-5273

57. Lonsdale, C. (1999). Effectively Managing Vertical Supply Relationships: A Risk Management Model for Outsourcing. Supply Chain Management: An International Journal, Vol. 4 No. 4, pp. 176-83, ISSN 1359-8546

58. Lowe, T.J., Wendell, R.E., & Hu, G. (2002). Screening Location Strategies to Reduce Exchange Rate Risk. European Journal of Operational Research, Vol.136, pp. 573-590, ISSN 0377-2217

59. Mol, M.J., Tulder, R.J.M., & Beije, P.R. (2005). Antecedents and Performance Consequences of International Outsourcing. International Business Review, Vol.14, pp. 599–617, ISSN 0969-5931

60. Monczka, R.M., & Trent, R.J. (1991). Global Sourcing: A Development Approach, International Journal of Purchasing and Materials Management, No. Fall, pp. 2-6.

61. Motwani, J., Youssef, M., Kathawala, Y., & Futch, E. (1999). Supplier Selection in Developing Countries: A Model Development. Integrated Manufacturing Systems, Vol.10, No.3,pp.154-161, ISSN 0957-6061

62. Munson, C.L., & Rosenblatt, M.J. (1997). The Impact Of Local Content Rules On Global Sourcing Decisions. Production and Operations Management, Vol.6, No.3, pp. 277-290, ISSN 1937-5956

63. Narasimhan, R., Talluri, S., & Mahapatra, S. K. (2006). Multiproduct, Multicriteria Model For Supplier Selection With Product Life-cycle Considerations. Decision Sciences, Vol. 37, No. 4, pp. 577-603, ISSN 1540-5915

64. Nooteboom,B. (1999). Inter-Firm Alliances : Analysis and Design. Routledge. ISBN 0-415-18154- 2, USA

65. Perron, S., Hansen, P. , Digabel, S., & Mladenovic, N. (2010). Exact And Heuristic Solutions Of The Global Supply Chain Problem With Transfer Pricing. European Journal of Operational Research, Vol. 202, No.3, pp. 864-879, ISSN 0377-2217

66. Petroni, A., & Braglia, M. (2000). Vendor Selection Using Principal Component Analysis. Journal of Supply Chain Management, Vol.36, No.2, pp. 63–69

67. Quinn, J.B. (1999). Strategic Outsourcing: Leveraging Knowledge Capabilities. Sloan Management Review, Vol. 40,No. 4, pp. 9-21

68. Quinn, J.B., & Hilmer, F.G. (1994). Strategic Outsourcing. Sloan Management Review, Vol. 35, No. 4, pp. 43-55

69. Rajan, A.J., Ganesh, K., & Narayanan, K.V. (2010). Application of Integer Linear Programming Model for Vendor Selection in a Two Stage Supply Chain.

70. International Conference on Industrial Engineering and Operations Management, Dhaka, Bangladesh, January 9-10, 2010.

71. Ramanathan, R. (2007). Supplier Selection Problem: Integrating DEA With The Approaches Of Total Cost Of Ownership And AHP. Supply Chain Management: An International Journal, Vol. 12, No.4, pp. 258-261, ISSN: 1359-8546

72. Ray, A., Sarkar, B., & Sanyal, S. (2008). A Holistic Approach For Production Outsourcing. Strategic Outsourcing: An International Journal, Vol.1, No.2, pp. 142-153, ISSN 1753-8297

73. Ren, J., Yusuf, Y. Y., & Burns, N. D. (2009). A Decision-Support Framework For Agile Enterprise Partnering. The International Journal of Advanced Manufacturing Technologies, Vol. 41, pp.180–192, ISSN

0268-3768

74. Saen, R.F. (2007). Suppliers Selection In The Presence Of Both Cardinal And Ordinal Data. European Journal of Operational Research, Vol. 183,pp. 741–747, ISSN: 0377-2217 Sarkis, J., & Talluri, S. (2002). A Model For Strategic Supplier Selection. Journal of Supply Chain Management. Vol.38, No.1, pp.18-28

75. Songhori, M.J., Tavana, M., Azadeh, A., & Khakbaz, M.H. (2011). A Supplier Selection And Order Allocation Model With Multiple Transportation Alternatives. International Journal of Advanced Manufacturing Technology, Vol.52, No.1-4, pp. 365–376, ISSN 0268-3768

76. Tam, F.Y., Moon, K.L., Ng, S.F., & Hui, C.L. (2007). Production Sourcing Strategies and Buyer-Supplier Relationships: A Study of the Differences Between Small and Large Enterprises in the Hong Kong Clothing Industry. Journal of Fashion Marketing and Management, Vol. 11, No.2, pp. 297-306, ISSN: 1361-2026

77. Tan, B. (2001). On Capacity Options In Lean Retailing. Harvard University, Center for Textile and Apparel Research, Research Paper Series. February.

78. Teng, S.G., & Jaramillo, H. (2005). A Model For Evaluation And Selection Of Suppliers In Global Textile And Apparel Supply Chains. International Journal of Physical Distribution & Logistics Management, Vol.35, No.7/8, pp. 503-523, ISSN 0960-0035.

79. Timmermans, K. (2005). The Secrets of Successful Low-Cost Country Sourcing. Accenture, Vol.2, pp. 62-72

80. Ting, S.C., & Cho, D.I. (2008). An Integrated Approach For Supplier Selection And Purchasing Decisions. Supply Chain Management: An International Journal, Vol. 13, No.2, pp. 116-127, ISSN : 1359-8546

81. Vidal, C.J., & Goetschalckx, M. (2001). A Global Supply Chain Model With Transfer Pricing And Transportation Cost Allocation. European Journal of Operational Research, Vol.129, pp. 134-158, ISSN 0377-2217

82. Vidal, C.J., & Goetschalckx, M. (2000). Modeling the Effect of Uncertainties on Global Logistics Systems. Journal of Business Logistics. Vol.21, No.1, pp. 95-120, ISSN 0197-6729

83. Vokurka, R., Choobineh, J., & Vadi, L. (1996). A Prototype Expert System For The Evaluation And Selection Of Potential Suppliers. International Journal of Operations & Production Management, Vol. 16, No.12, pp. 106-127, ISSN: 0144-3577

84. Wang, G., Huang, S.H., & Dismukes, J.P. (2004). Product-Driven Supply Chain Selection Using Integrated Multi-Criteria Decision-Making Methodology. International Journal of Production Economics, Vol. 91, pp. 1-15, ISSN: 0925-5273

85. Wang, H.S., & Che, Z.H. (2007). An Integrated Model For Supplier Selection Decisions In Configuration Changes. Expert Systems with Applications, Vol.32, pp. 1132–1140, ISSN 0957-4174

86. Wang, H.S. (2008). Configuration Change Assessment: Genetic Optimization Approach With Fuzzy Multiple Criteria For Part Supplier Selection Decisions. Expert Systems with Applications, Vol. 34, pp. 1541-1555, ISSN: 0957-4174

87. Wang, J.J, Hu, R.B., & Diao, X.J. (2008-a). Developing an Outsourcing Decision Model Based on ELECTREI Method. 4th International Conference on Wireless communications, Networking and Mobile Computing, pp. 1-4, ISBN 978-1-4244-2107-7, Dalian, October12-14, 2008.

88. Wang, J.J., Gu, R., & Diao, X.J. (2008-b). Using A Hybrid Multi-Criteria Decision Aid Method For Outsourcing Vendor Selection. 4th International Conference on Wireless communications, Networking and Mobile Computing, pp. 1-4, ISBN 978-1-4244-2107-7, Dalian, October 12-14, 2008.

89. Weber, C.A., Current, J.R., & Benton, W.C. (1991). Vendor Selection Criteria And Methods. European Journal of Operational Research, Vol. 50, pp. 2-18,I SSN 0377-2217

90. Wei, S., Zhang, J., & Li, Z. (1997). A Supplier Selecting System Using A Neural Network, IEEE International Conference on Intelligent Processing Systems, pp. 468-471, ISBN: 0-7803-4253-4, Beijing China, October 28-31, 1997.

91. Wu, D. (2009). Supplier selection: A Hybrid Model Using DEA, Decision Tree And Neural Network. Expert Systems with Applications, Vol.36, pp. 9105-9112, ISSN: 0957-4174

92. Xia, W., & Wu, Z. (2007). Supplier selection with multiple criteria in volume discount Environments. The International Journal of Management Science, Vol. 35, pp. 494-504,ISSN: 0305-0483

93. Yang, S., Yang, J., & Abdel-Malek, L. (2007). Sourcing With Random Yields And Stochastic Demand: A Newsvendor Approach. Computers & Operations Research, Vol.34,pp.3682 – 3690, ISSN: 0305-0548

94. Zeng, A.Z. (2000). A Synthetic Study of Sourcing Strategies. Industrial Management & Data Systems, Vol.100, No.5, pp.219-226, ISSN: 0263-5577

Chapter 7

PLATFORM FOR INTELLIGENT MANUFACTURING SYSTEMS WITH ELEMENTS OF KNOWLEDGE DISCOVERY

Tomasz Mączka and Tomasz Żabiński

Rzeszów University of Technology Poland

INTRODUCTION

Numerous and significant challenges are currently being faced by manufacturing companies. Product customization demands are constantly growing, customers are expecting shorter delivery times, lower prices, smaller production batches and higher quality. These factors result in significant increase in complexity of production processes and the necessity for continuous optimization. In order to fulfil market demands, managing the production processes require effective support from computer systems and continuous monitoring of manufacturing resources, e.g. machines and employees. In order to provide reliable and accurate data for factory management personnel the computer systems should be integrated with production resources located on the factory floor. Currently, most production systems are characterized by centralized solutions in organizational and software fields. These systems are no longer appropriate, as they are adapted to high volume, low variety and low flexibility production processes. In order to fulfil current demands, enterprises should reduce batch sizes, delivery times, and product life-cycles and increase product variety. In traditional manufacturing systems this would create an unacceptable decrease in efficiency due to high replacements costs, for example. (Christo & Cardeira, 2007) Modern computer systems devoted to manufacturing must be scalable, reconfigurable, expandable and open in the structure. The systems should enable an on-line monitoring, control and maximization of the total use of manufacturing resources as well as support human interactions with the system, especially on the factory floor. Due to vast amounts of data collected by the systems, they should automatically process data about the manufacturing processes, human operators, equipment and material requirements as well as discover valuable knowledge for the factory's management personnel. The new generation of manufacturing systems which utilizes artificial intelligence techniques for data analyses is referred to as Intelligent Manufacturing Systems (IMS). IMS industrial implementation requires computer and factory automation systems characterized by a distributed structure, direct communication with manufacturing resources and the application of sophisticated embedded devices on a factory floor. (Oztemel, 2010) Many concepts in the field of organizational structures for manufacturing have been proposed in recent years to make IMS a reality. It seems that the most promising concepts are: holonic (HMS or Holonic Manufacturing System), fractal, and bionic manufacturing. Further references can be found in (Christo & Cardeira, 2007). In general, it could be stated that a promising organizational structure is a conglomerate of distributed and autonomous units which operate as a set of cooperating entities. It would be impossible to successfully implement the new organizational concepts in the manufacturing industry without suitable distributed control and monitoring hardware and software.

In publications (Leitão, 2008), (Colombo et al., 2006) an agent-based software is designated as technology for industrial IMS realization, regardless of the chosen organizational paradigm. The integration of HMS and multi-agent software technology is currently presented as the most promising foundation for industrial implementations of IMS. The HMS paradigm is based on concepts originally developed by Arthur Koestler in 1969 with reference to social organizations and living organisms. The term holon describes a basic unit of organization and has two important characteristics: autonomy and co-operation. In a manufacturing system, a holon can represent a physical or logical activity, e.g. a machine, robot, order, machine section, flexible manufacturing system and even a human operator. The holon possesses the knowledge about itself and about the environment and has an ability to cooperate with other holons. From this viewpoint, a manufacturing system is a holarchy, which is defined as a system of holons organized in a dynamical hierarchical structure. Manufacturing system goals are achieved by cooperation between holons. Due to conceptual similarities of HMS and agent-based software, it seems clear that their combination should create a promising platform for an industrial IMS implementation. On the other hand, humans still play an important role in manufacturing systems and in spite of predictions from the seventies, which suggested that human operators would no longer be needed in fully automated production, they even play a more important role nowadays than they did in the past (Oborski, 2004). Proper cooperation between humans and machines or humans and manufacturing control systems can significantly improve overall production effectiveness, so human system interface design is still an active research area. Human System Interfaces (HSI) are responsible for efficient cooperation between operators and computer systems. (Gong, 2009) It seems clear that convenient and reliable human system interaction is a key factor for successful industrial IMS implementation. In this chapter, results of the project devoted to the development of a hardware and software platform for IMS are to be described. The platform based on Programmable Automation Controllers (PAC) has already been successfully tested by being included in everyday production processes in four small and medium Polish metal component production companies. The platform was employed in order to monitor production resources in real-time and to conduct communication between computer systems, machines and operators. On the basis of the tests results, it has been experimentally proven that the main development-related barrier for real deployment of IMS (Leitão, 2008), i.e. absence of industrial controllers with appropriate capabilities, is out-of-date. Within 28 months of the system operation it has been proven that modern PAC are capable of running data processing, communication and graphical user interface modules directly on the PAC controller in parallel with PLC programs. The hardware and software system which has been created constitutes of a platform for future complete implementation of the IMS.

The project has been created by the Department of Computer and Control Engineering, Rzeszów University of Technology, in cooperation with Bernacki Industrial Services company and the students from the Automation and Robotics scientific circle called ROBO (ROBO, 2011). The project has been made under the auspices of Green Forge Innovation Cluster. The long-term goal of this project is the development and industrial implementation of full IMS concept for small and medium production companies. The goal determines two main assumptions for the selection of hardware and software elements. The first assumption is a possibility to include in the system newer machines with advanced control equipment as well as older ones without controllers. The second one is a reasonable cost of the system. In the paper (Żabiński et al., 2009) the results of the first stage of the project were presented. During the first stage, the hardware was selected and the prototype and functionality limited testbed for one machine was installed in a screw manufacturing factory. The system was tested in a real production process. The first stage of the project was to provide valuable benefits for the factory management board in the field of monitoring availability, performance and quality of operation of the machines and operators. Additionally, control tasks in a PLC layer of the system were done using ST (Structured Text) programming language for the machines which had not been previously equipped with a controller. The success of the first stage resulted in the project continuation and the findings of the current stage are presented in this paper and in (Mączka & Czech, 2010), (Żabiński & Mączka, 2011), (Mączka & Żabiński 2011), (Mączka et al., 2010). The main goal of the current project stage is to develop and test in a real production environment a system structure with PACs integrated with touchable panels for each machine. Some improvements of HSI for the factory floor are being made, according to suggestions acquired during the first stage, and new functionalities for different organizational units, i.e. maintenance, tool and material departments. The important part of current and future works is to employ artificial intelligence and data mining technology to give factory management personnel reliable and long-term knowledge about the production processes.

SYSTEMS TESTBED STRUCTURES

Up to now, four system industrial testbeds have been constructed. The first one was installed in a screw manufacturing company which is a member of the Green Forge Innovation Cluster. The Green Forge Innovation Cluster is an association of metal production companies and scientific institutions from southern Poland, which aims at innovative solutions development for metal components production. The second one was installed at the WSK PZL-Rzeszów company, in the department which produces major rotating parts for the aviation industry. The first testbed consists of two machine sections formed by cold forging press machines. The first section includes machines without PLC controllers but the second one consists of 12 modern machines equipped with PLCs and advanced cold forge process monitoring devices. The second testbed includes one production line with four CNC vertical turning lathes and two CNC machining centres. The machines in the testbeds have been operated by experienced operators who interact with the system using various peripheral devices, such as barcode readers, electronic calipers and industrial touch panels. (Żabiński & Mączka, 2011) There is also a third testbed installed in a different department at "WSK PZL-Rzeszów", which monitors one machine.

During the project, a mobile testbed with GSM communication was also constructed. The purpose of a mobile testbed is to allow production companies to test the system without bearing the costs of communication infrastructure installation. Thanks to such a testbed, companies can better define system functionalities better which are very important for them, taking into account the production profile and organizational structure. Currently, this testbed has been installed in a metal component manufacturing company, where one machine has been monitored. (Mączka & Żabiński, 2011) In the hardware and software part of the system, three main layers can be distinguished: a factory floor hardware and software, a data server layer and WWW (World Wide Web) client stations.

Factory Floor Layer

Due to the scalable, reconfigurable, expandable and open structure of the platform, the industrial implementations differ in functionality as well as in hardware and software elements installed on a factory floor.

Testbed with PACs

In the first implementation type, Programmable Automation Controllers (PACs), also known as embedded PCs, are used on a factory floor. PACs are equipped with operating systems and meet the demands of modern manufacturing systems as they combine features of traditional Programmable Logic Controllers (PLCs) and personal computers (PCs). The main feature of PACs is the ability to use the same device for various tasks simultaneously, e.g. data acquiring, processing and collection; process and motion control; communication with databases or information systems; Graphical User Interface (GUI) implementation, etc. There are two kinds of Windows system available for the controllers, i.e. Windows CE and Windows XP Embedded. Windows CE is equipped with .Net Compact Framework, Windows XP Embedded is equipped with .Net Framework. There are benefits of using the XP Embedded platform, e.g. homogeneity of the software platform for controllers and PC stations as well as availability of network and virus protection software. Due to the financial reasons, Ethernet network for communication and controllers with Windows CE were chosen for the two testbeds. In the system, PACs acquire data form machines using distributed EtherCAT (Granados, 2006) communication devices equipped with digital or analog inputs. Each PAC is equipped with Windows CE 6.0 operating system, real-time PLC subsystem, UPS (Uninterruptible Power Supply), Ethernet as well as RS-232/485 interfaces for communication and DVI (Digital Visual Interface)/USB (Universal Serial Bus) interfaces for touchable monitors connection. One controller with peripherals, i.e. an industrial 15" touch panel, RFID (Radio-frequency Identification) cards reader and barcode reader, is installed in each machine section or production line. The hardware system structure for a factory floor is shown in Fig. 1. The software for PACs consist of two layers. The first layer is PLC software written in ST (Structured Text) language, which is mainly responsible for reading and writing physical inputs and outputs. The second layer constitutes the application written in C# language for .NET Compact Framework (CF), which runs under Windows CE. (Microsoft Developer Network, 2011)

- the module for communication between the PLC program and other system parts,
- the module for communication with database using web services technology,
- the operator's GUI.

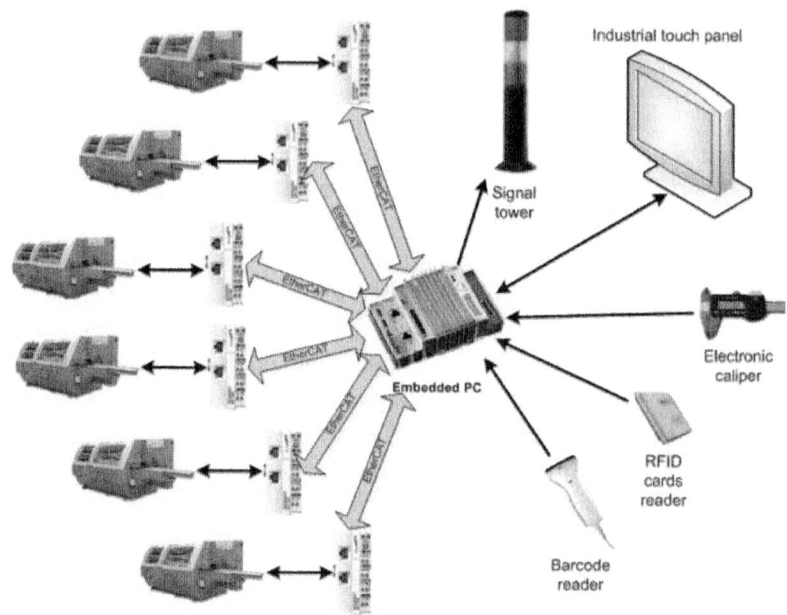

Figure. 1: Hardware structure of the platform for the first implementation type

PLC control programs run in the PLC layer on the same device simultaneously with GUI, data processing and database communication modules which run in Windows CE layer (Fig. 2). The ADS (Automation Device Specification) protocol enables C# programs to read and write data directly from and to PLC programs via names of PLC variables. It significantly simplifies the communication between PLC and C# applications.

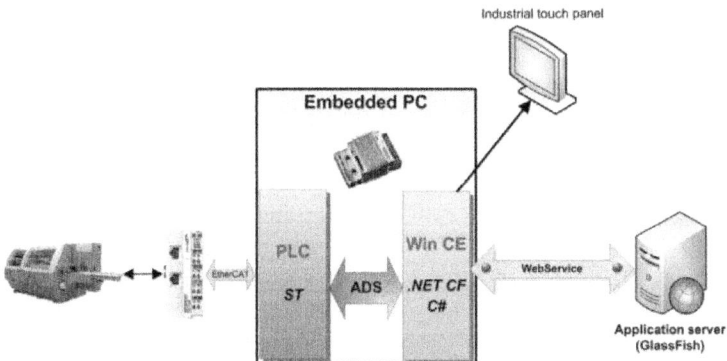

Figure. 2: PAC software structure

The PLC and Windows software for embedded PC controller was designed and implemented in order to control up to 6 machines. It gives flexibility in the system structure, e.g. for more demanding PLC or CNC control tasks there is a configuration of one controller for one possible machine. For simple machines or machines already equipped with controllers, it is possible to create a configuration with one embedded PC controller and up to six machines. This configuration can be used, for instance, to incorporate machine sections into the system. Currently, the new implementation type supported by EU funds, is under construction. It concerns installation of PACs integrated with touchable panels for each machine (Fig. 3). The panel allows operators to interact with the system and input basic data regarding corresponding machines, e.g. reason for production stoppage and references for orders, materials and tools. In this case, the computational power of each PAC is devoted to one machine and is used for data gathering, intelligent data analysis, intelligent condition monitoring, alarms detection, etc. The PACs models were selected in order to provide sufficient performance for the future multi-agent system structure version, with separate machine agents running on each PAC. In this implementation type, apart from separate PAC for each machine, personal computers with HSI for operators can be flexibly connected to the system. PCs are treated as additional system "access points", which allow interaction with the system on the same level as PACs. Additionally, it can e.g. provide technical documentation for a particular production process in a more convenient way than on the PAC with a small touchpanel.

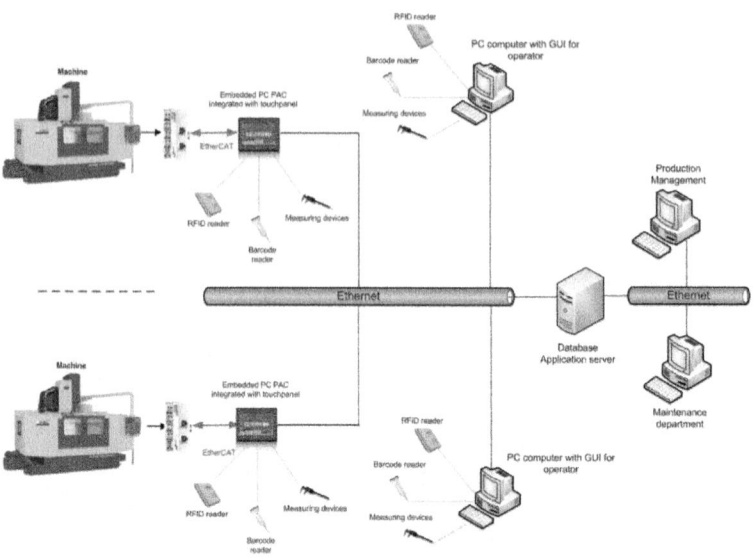

Figure. 3: Hardware structure of the platform with PAC for single machine

Testbed with Industrial PC

System structure in the "WSK PZL-Rzeszów" testbed is going to be modified, in order to obtain more flexible architecture. The new structure should also simplify the system implementation in companies with various production and data resources. The new software structure should enable its easy adaptation to the needs of other production sectors. During the development of the testbed, additional diagnostics and process monitoring equipment will be included in the system, e.g. quality measurement devices, current and force sensors, etc. In concern to the WSK testbed, it is planned to include an additional 64 machines in the system. Due to customer demands, one industrial PC computer will be installed for each production line. The industrial PC's task is monitoring states of the machines included in particular production line on base of digital (machine work mode, machine engine state) and analog (spindle load) signals from machines' control systems. Windows XP Embedded with real-time and soft PLC TwinCAT subsystem is the operating environment for industrial PC; connection with input/output modules is performed via EtherCAT bus using star network topology.

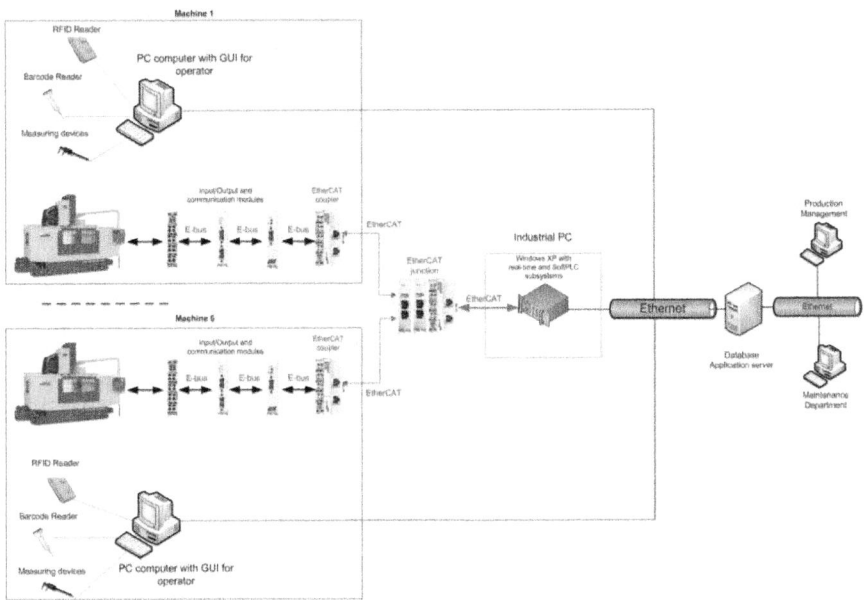

Figure. 4: Hardware structure with industrial PC

Chosen topology minimizes communication problems if some part of the EtherCAT network infrastructure will fail, e.g. in case of network cable break. In this type of implementation, machine operators interact with the system using PCs workstations, placed near monitored machines. The basic scope of data, which can be viewed and input carried out by operators, is similar like in PAC implementation. Detailed description will be given in section 3.1. Additionally, the system is going to be integrated with an SAP business software and will be used for delivering electronic versions of technical and quality control documents directly to operators' workstations. The system should support production management as well as support the maintenance department by the usage of advanced and intelligent machine condition monitoring software tools. The new testbed structure is shown in Fig. 4. (Żabiński & Mączka 2011)

Data Server and end Clients Layer

The data and application server layer is common for all the structures. It includes the PostgreSQL database server and the GlashFish application server. GlassFish is an opensource application server compatible with Java Platform Enterprise Edition (J2EE). (GlassFish Community, 2011) PostgreSQL is open source object-relational database system which conforms to the ANSI-SQL (Structured Query Language) 2008 standard. (PostgreSQL, 2011) The application server hosts communication and data processing modules with web services written in Java. The WWW client layer utilizes websites written in JSF (Java Server Faces), JSP (Java Server Pages), AJAX (Asynchronous Javascript and XML) and also works under GlassFish server control. Communication between PACs or PCs and the database and between the presentation layer and the database is performed using web services or Enterprise Java Beans (EJB) technology.

HUMAN SYSTEM INTERFACES

Human System Interface, which was developed for the IMS project, consists of two main layers, i.e. a factory floor layer and a WWW layer. The first layer is a GUI application which runs on an embedded PC installed on the factory floor. In this layer the communication between an operator and the system is done via a 15" touchable monitor. The second layer is a web page accessed through a web browser from the factory intranet or the Internet. The Polish language is used in HSI, as the system has been installed in Polish factories. Due to this reason, the GUI language presented in this section is Polish.

Factory Floor Layer

The HSI for factory floor layer has two main operation modes, i.e. locked and unlocked. In the locked mode an operator can only observe information presented on the screen. In the unlocked mode an operator can interact with the system. An operator can change the HSI mode using his RFID card. Thanks to the RFID operator's badges a security policy was implemented on the factory floor. In the locked mode visual information of machines operation mode, production plan and plan realization and the necessity of an operator interaction with the system is presented. The necessity of an operator interaction with the system is indicated by the blinking of a panel associated with the machine which needs intervention. The locked mode screen for a machine section with six machines is presented in Fig. 5. In the unlocked HSI mode an operator can perform various tasks connected with the system, e.g. login, logout, taking up shift, order selection or confirmation, stop reason and quality control data input etc. The unlocked mode screen for a machine section with six machines is presented in Fig. 6. The HSI consists of two main sections, i.e. the system section (it allows login, logout, etc.) and the machines section (it allows data input for particular machines). Small rectangles with letters O, SR, QC, S, associated with each machine panel (a large rectangle with machine name, e.g. Tłocznia T-19), indicates the action which should be performed for the particular machine, i.e. O – order selection or confirmation, SR – stop reason input, QC – quality control data input and S – service.

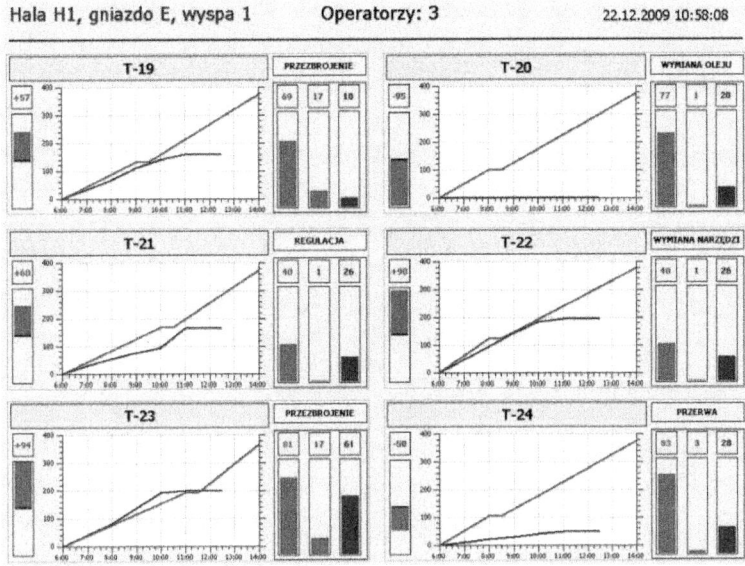

Figure. 5: The factory floor HSI locked mode screen for six machines

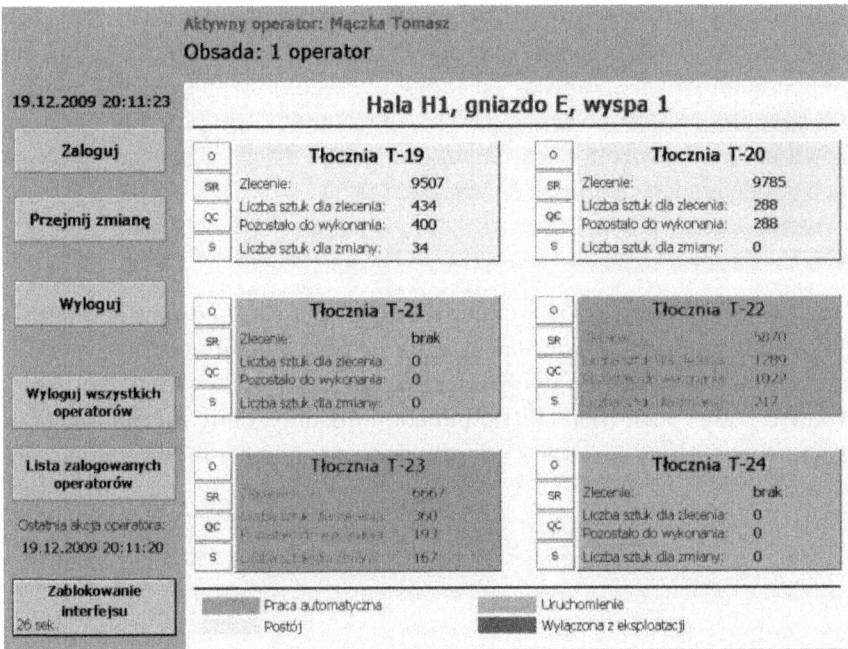

Figure. 6: The factory floor HSI unlocked mode screen for six machines

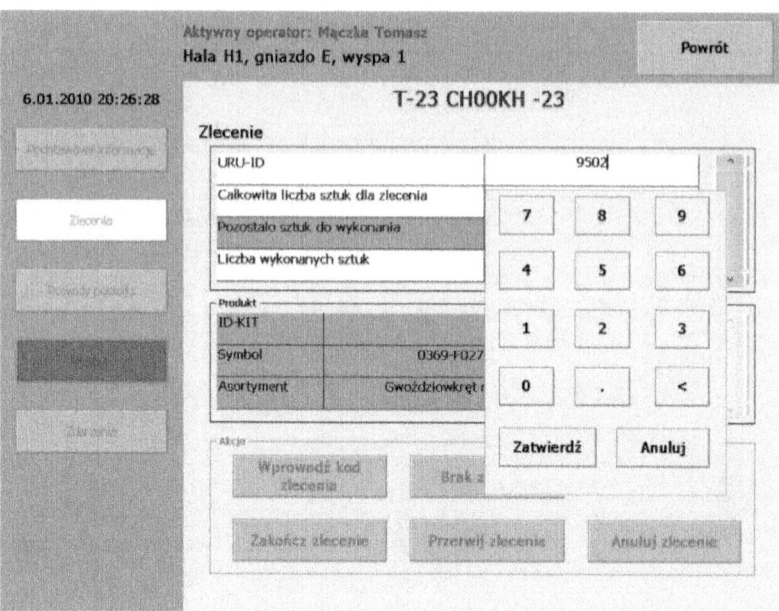

Figure. 7: Order inputting screen for particular machine

When the machine panel is blinking, an operator can quickly determine the operation which should be performed, the color of the letters O, QC, SR or S becomes red for the active action. An example of an order input screen is shown in Fig. 7. Operators can input order code using on-screen keyboard or via barcode reader. (Żabiński & Mączka, 2011)

WWW Layer

The WWW layer includes two main modules, i.e. an on-line view and statistics. The on-line view enables on-line monitoring of machines operation mode, e.g. production, stoppage, lack of operator and also other information like: operator identifier, order identifier, shift production quantity, daily machine operation structure or detailed history of events. The on-line view for a production hall is presented in Fig. 8.

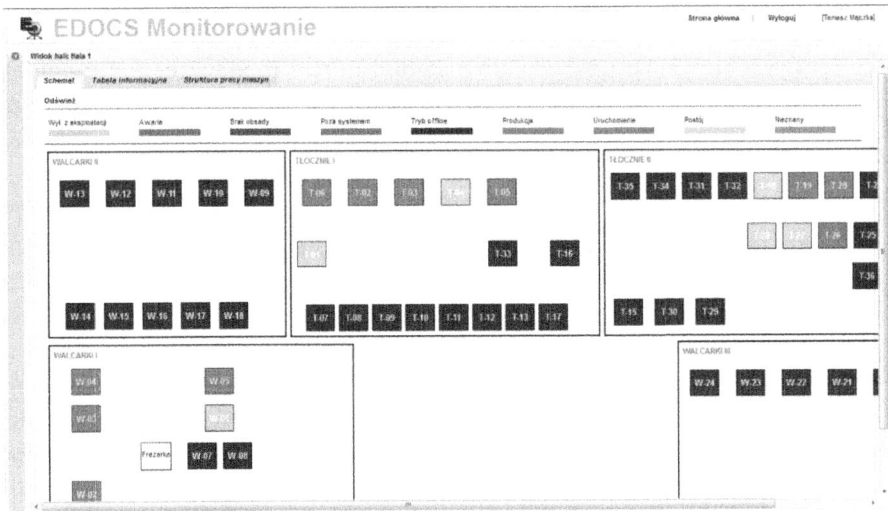

Figure. 8: Production state on-line view for a production hall

The statistics module enables computing some statistical factors concerning: machines work time, production quantity, failures, orders, operators work time etc. It enables users to configure statistics parameters, i.e. analysis time interval, statistics elements, type of presented data (for instance daily or weekly analysis type) and chart type (e.g. bar graph, line graph). An example of a statistics screen for machine production quantity with its configuration options is shown in Fig. 9. (Żabiński & Mączka, 2011)

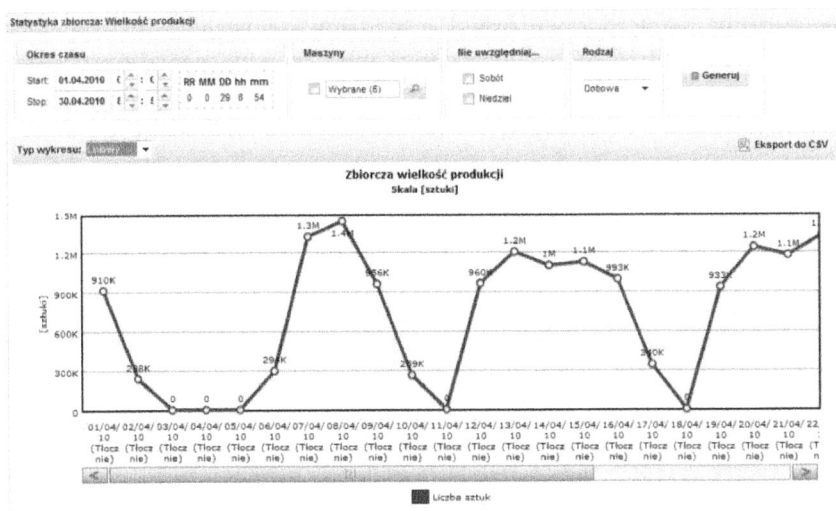

Figure. 9: Machines production quantity statistic with its configuration options

SYSTEM OPERATION RESULTS

Currently, four system testbeds are installed in real production environments in small and medium factories and have been used in daily production processes. All of the testbeds currently utilize the first structure described in section 2.1.1, with embedded PCs on the factory floors. The first testbed has been installed in a screw manufacturing factory since 16 May, 2009. Eighteen machines for cold forging are currently monitored. The second testbed was installed at the WSK PZL-Rzeszów company, in the department which produces major rotating parts for the aviation industry and it has been in operation since 21 September, 2010 with six CNC machines included in the system. The new system implementation with the second type structure for this testbed, described in section 2.1.2, is under construction. There is also one testbed where a single machine is monitored using a mobile (all-in-one) system testbed with GSM communication. The next separated testbed with one machine was installed in April 2011 in a different department at "WSK PZL-Rzeszów"of the aviation parts producing company. Currently, the system is responsible for collecting data concerning machine operation and operators' work. The PLC layer is responsible for detecting and registering events which occurred in the machines, for instance the oil pump and the main motor start/ stop, failure and emergency signals, the machine operational mode (manual or automatic), signals from diagnostics modules (process monitoring devices), etc. The PLC program is also responsible for registering the quantity produced. Information about events, including timestamps, machine and operator identifiers and other additional parameters, is stored in the database. Two mechanisms are used to store data in the database, such as: an asynchronous event driven method and a synchronous one with a 10 second time period for diagnostics purposes. The system is also responsible for detecting and storing information on breakdowns, setup and adjustments, minor stoppages, reduced speed etc. Every production stoppage must be assigned with an appropriate reason. Tool failure, for example, are automatically detected, while others have to be manually chosen by operators via the HSI. On the server side, there are software modules used for calculating different KPIs (Key Performance Indicators), e.g. production efficiency, equipment and operators efficiency etc. The real production quantity report as a function of day, calculated for time interval from 1- 03-2011 to 31-03-2011 for 6 machines, is presented in Fig. 10. During this period, the planned production time for the machines was 24 hours per day (3 shifts). As shown in Fig. 10, there were some fluctuations of production quantity.

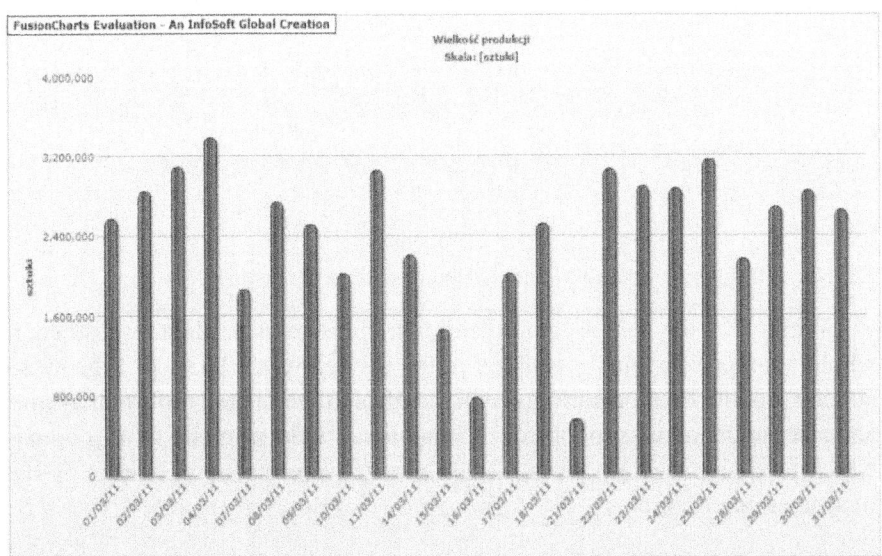

Figure. 10: Production quantity report – number of produced items as a function of days

A machine operation time structure is analyzed and can be shown as a horizontal graph (Fig. 11). At the moment there is a possibility of analyzing data from three points of view: a general view, a view with stop reasons and a detailed view. The general view divides machine operational time into three categories, i.e. the operator's absence, the automatic production and the stoppage. In the analysis for the view with stop reasons, each stoppage time period is associated with the appropriate stop reason. In the detailed view, periods of the manual machine operation are distinguished in each stoppage time. Different colors are designated for appropriate time intervals (Fig. 11), e.g. the stoppage – light brown, the automatic production – green, the manual operation – dark green, the start-up time – blue, the electrical breakdown – red, etc. During the long term test, when the system was included in the regular daily production, it was proven that the selected hardware and software platform is suitable for industrial implementation of the IMS. The software modules (HSI, communication, web services, data acquisition) for Windows CE have been running successfully on the embedded PC controller in parallel to PLC programs. (Mączka & Czech, 2010)

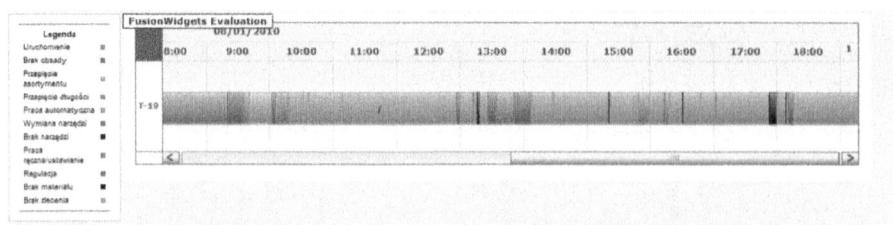

Figure. 11: Machine operation time structure with stop reasons

Within 704 days of system operation in the screws production testbed, the whole number of events registered in the testbed was 3,883,374. The system covers: production monitoring, quality control, material and tools management and also fundamental support for the maintenance department. In the company which produces aviation parts, 57,230 events were registered within 59 days of system operation.

ASSOCIATION RULES APPLICATION FOR KNOWLEDGE DISCOVERY

Taking into account the limited number of machines included in the currently working system testbeds, it can be stated that the number of data collected is considerable. A longterm data analysis to make reasoning and generalized conclusions about production processes would be a demanding task for human analysts. Therefore, there is a need to employ artificial intelligence and data mining technology to give factory management personnel reliable knowledge of the production processes. In this section, the results of the initial tests in the areas of applying data mining and artificial intelligence techniques to discover knowledge about the production processes, are to be described. It is expected that continuously discovered knowledge will support everyday production process management and control, thus providing the answers to numerous questions, e.g. what are the bottlenecks in the production systems etc. Moreover, it is envisaged that the system will be able to automatically identify relationships in the production systems, discover possibilities for more effective usage of production resources and use Statistical Process Control (SPC) with artificial intelligence support for early detection of possible problems in production systems. Initial work in this area concerns the creation of tools for automatic rules (patterns) generation, which will describe relations between values in the database. The discovered patterns could be used for detecting operators' improper actions, which could have an influence on machine operation, e.g. increasing downtime duration and number of breakdowns. The rules are referred to as association rules.

Introduction to the Experiment

The goal of association rules is to detect relationships between specific values of categorical variables in large data sets.

The formal definition of the problem is as follows: Let $D = \{t_1; t_2; ...; t_m\}$ be a set of m transactions, called data set or database. Let $I = \{i_1; i_2; ...; i_n\}$ be a set of possible n binary attributes for transaction, called items. Single transaction T is a set of items such that $T \subseteq I$. Assume that X is a set of some items from I, so $X \subseteq I$. A transaction T contains X if the transaction contains all items from X, so $X \subseteq T$.

Table 1: Structure of raw events extracted from production database

plc_time	event_type_id	machine_id
2011-01-04 07:59:25.784	2000	41
2011-01-04 08:26:27.565	2001	41
2011-01-04 08:28:49.845	2000	41
2011-01-04 08:29:42.705	2001	41
2011-06-14 21:49:38.032	2000	44
2011-06-14 23:29:17.732	2001	44
2011-07-11 06:59:49.812	2000	68
2011-07-11 07:35:15.332	2001	68
2011-07-11 07:41:44.872	2000	68
2011-07-11 08:21:07.812	2001	68

An association rule is an implication of the form $X \Rightarrow Y$, where $X \subset I$, $Y \subset I$, and $X \cap Y = \emptyset$. The rule $X \Rightarrow Y$ holds in the transaction set D with confidence c if c% of transactions in D that contain X also contain Y. The rule $X \Rightarrow Y$ has support s in the transaction set D if s% of transactions in D contain $X \cup Y$. Given a set of transactions D, the problem of mining association rules is to generate all association rules that have support and confidence greater than the user-specified minimum support (minsup) and minimum confidence (minconf) respectively. (Agrawal & Srikant, 1994). The experiment of finding association rules in production data has been performed. The first step of the experiment was to choose the subset of data to analyze, i.e. time period and attributes, and to extract raw data from screw manufacturing company database to CSV (Comma Separated Values) text format. CSV format was chosen because of the possibility of loading data directly into data mining software, i.e. Statistica or Weka. 127232 events registered on 12 machines from 4.01.2011 to 16.07.2011, concerning machine operational state, were extracted using SQL query and pgAdmin database management tool. The structure of events is shown in Table 1, the table contains only a subset of registered events.

Attributes of a single event are:

- plc_time– time of event registration in the PLC layer
- event_type_id– type of registered event, 2000 is production start, 2001 is production stop
- machine_id– identifier of a machine for which the event was registered

After consultation with the company production management personnel, an assumption has been made that length of times of continuous machine state, i.e. length of production and length of stoppage will be important factor to analyze. It seems to be clear that if continuous production time of a particular machine lasts longer than the others, this machine works more efficiently, without the need for operator action. It is worthwhile noticing that lower number of stoppages should positively affect machine lifetime and save energy.

Data Preparation

The format for raw data presentation in Table 1 is not useful for discovering associations concerning machines continuous interval length, as raw events do not directly reflect particular machine state during a particular period of time. Because of this, data needs to be pre-processed to the list of production and stoppage intervals for each machine. Each record should contain interval start date, interval end date and interval length. Pre-processing task was done using Python script, which analyzes events list and produces stoppage intervals, if the current analyzed event is 2001 (production stop) and next analyzed event is 2000 (production start). In the opposite situation, production interval is inserted to the result list. Data structure after the pre-processing phase is shown in Table 2.

Table 2: Data structure after the pre-processing phase

start	End	type	length_min	machine_id
2011-01-04 07:59:25.784	2011-01-04 08:26:27.565	W	27.03	41
2011-01-04 08:26:27.565	2011-01-04 08:28:49.845	S	2.37	41
2011-01-04 08:28:49.845	2011-01-04 08:29:42.705	W	0.88	41
...				
2011-06-14 21:49:38.032	2011-06-14 23:29:17.732	W	99.66	44
...				
2011-07-11 06:59:49.812	2011-07-11 07:35:15.332	W	35.43	68
2011-07-11 07:35:15.332	2011-07-11 07:41:44.872	S	6.49	68
2011-07-11 07:41:44.872	2011-07-11 08:21:07.812	W	39.38	68

Attributes of single interval:

- start – timestamp of interval begin,
- end – timestamp of interval end,
- type – interval type, W is work, S is stoppage,
- len_min – interval time length in minutes,
- machine_id – identifier of machine associated with interval.

Start and end timestamps has only informational role and they are omitted in the process of finding associated rules. However, data presented in Table 2 are not ready for application of associated rules finding algorithms. It results from fact, that known association rules discovering algorithms deals with data, whose attributes have discrete or categorical values. In above case, attributes type and machine_id are categorical, but len_min has continuous values. The solution of this problem is mapping attributes with continuous values to categorical attributes, referred in (Agrawal & Srikant, 1996) as partitioning quantitative attributes. This transformation was, like the previous, performed by Python script. Number of categories and length of each category's interval were chosen based on minimum and maximum values of len_min, in order to obtain regular distribution of data in generated categories. 46 categories were generated, starting from [0-0.1m] (interval length greater than 0 to 0.1 minutes, or 10 seconds) to [700-+Inf m] (interval length greater of equal 700 minutes). Example values of processed items are contained in table 3. Letter 'M' was added before machines identifiers to indicate that this is categorical, not numerical attribute.

Table 3: Data after partitioning quantitive attribute len_min to categorical attribute length_category

start	End	type	length_category	machine_id
2011-01-04 07:59:25.784	2011-01-04 08:26:27.565	W	[20.1-30.1m]	M41
2011-01-04 08:26:27.565	2011-01-04 08:28:49.845	S	[2.1-3.1m]	M41
2011-01-04 08:28:49.845	2011-01-04 08:29:42.705	W	[0.8-0.9m]	M41
...				
2011-06-14 21:49:38.032	2011-06-14 23:29:17.732	W	[90.1-100.1m]	M44
...				
2011-07-11 06:59:49.812	2011-07-11 07:35:15.332	W	[30.1-40.1m]	M68
2011-07-11 07:35:15.332	2011-07-11 07:41:44.872	S	[6.1-7.1m]	M68
2011-07-11 07:41:44.872	2011-07-11 08:21:07.812	W	[30.1-40.1m]	M68

Finding Association rules using Weka

For finding associations in previously prepared data, Weka (Waikato Environment for Knowledge Analysis) software was used. This is a popular suite of a machine learning software written in Java that was developed at the University of Waikato, which is available under the GNU General Public License. Weka is a collection of machine learning algorithms for data mining tasks. Weka contains tools for not only for association rules, but also for classification, regression, clustering and visualization. Its architecture is well-suited for developing new machine learning schemes. (Hall et al., 2009) In the first step, Weka knowledge explorer was run and preprocessed mining dataset was loaded from CSV file. Weka displays list of attributes in dataset and its basic information like categories, number of items in each category etc. (Fig. 12).

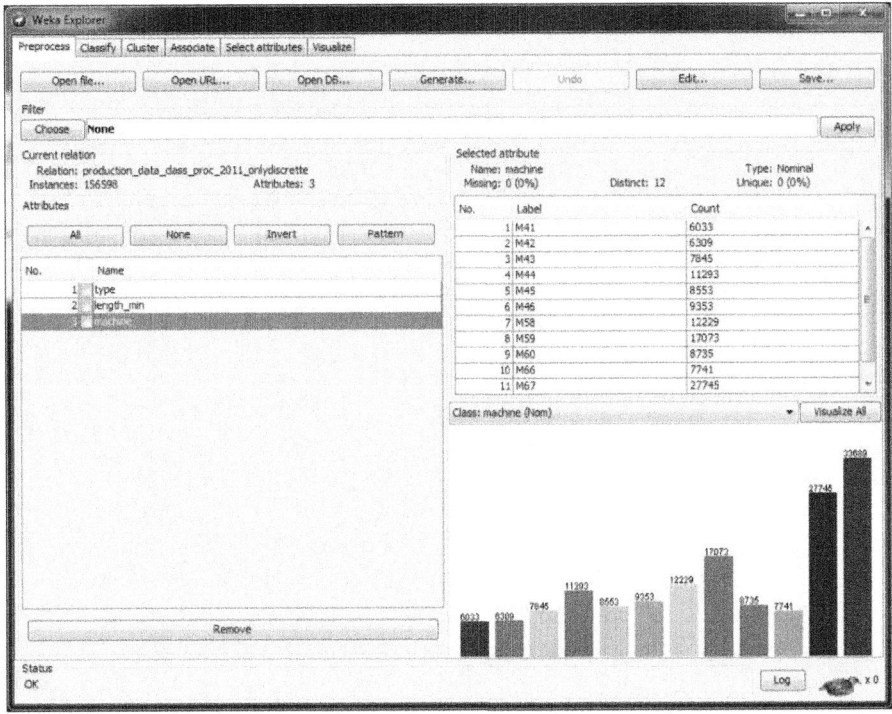

Figure. 12: Weka initial screen after data loading

The next step was running module for association rules generation, choosing algorithm for mining rules and configuring algorithm's parameters. Weka provides few rules discovering algorithms, including Apriori, Predictive Apriori, Generalized Sequential Patterns. In the described experiment, most popular Apriori algorithm was selected with parameters: support 0.1 (10%), confidence 0.3 (30%). The small value of support was choosen experimentally to not omit rules for particular machines, and purpose of 30% confidence was to ignore irrelevant rules, which applies only to relative small part of data set. Configuration parameters are shown in Fig. 13.

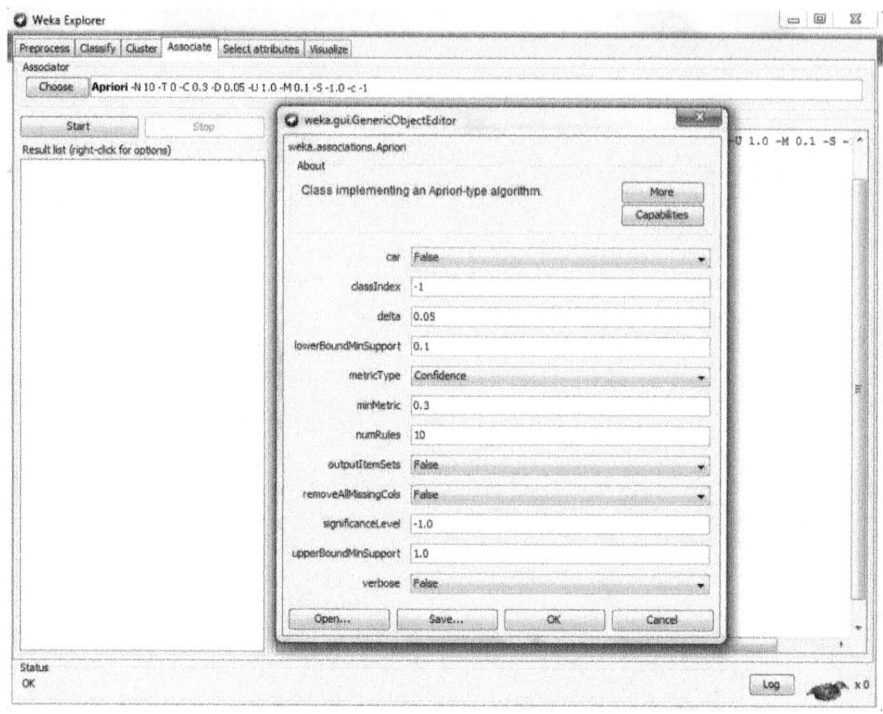

Figure. 13: Apriori algorithm configuration

The algorithm found 10 rules listed below:

1. <u>machine=M67 [27745] ==> length_min=0-0.1m [20099] conf:(0.72)</u>
2. <u>machine=M68 [33689] ==> length_min=0-0.1m [23825] conf:(0.71)</u>
3. length_min=0-0.1m [64730] ==> type=W [37527] conf:(0.58)
4. machine=M68 [33689] ==> type=W [16845] conf:(0.5)
5. machine=M68 [33689] ==> type=S [16844] conf:(0.5)
6. <u>type=W [78305] ==> length_min=0-0.1m [37527] conf:(0.48)</u>
7. length_min=0-0.1m [64730] ==> type=S [27203] conf:(0.42)
8. length_min=0-0.1m [64730] ==> machine=M68 [23825] conf:(0.37)
9. <u>type=S [78293] ==> length_min=0-0.1m [27203] conf:(0.35)</u>
10. length_min=0-0.1m [64730] ==> machine=M67 [20099] conf:(0.31)

Number of records which contain attributes values listed in predecessor and consequent are present in bracket squares. Some potentially interesting rules are underlined, their interpretation may be as follows. Rules 1 and 2 indicates that for machine M67 and M68, production or stoppage interval lasts usually relatively short – from above 0 to 10 seconds. It suggests that listed machines may have some troubles with stable work, which can lead to low efficiency. Potential reasons of this situation are mechanical problems, improper material, improper operator's actions etc. This situation is also covered by rules 8 and 10, where predecessor and consequent are reversed, these rules are more general. On the other hand, 48% of work intervals and 35% of stoppage intervals lasts up to 10 seconds, so short time intervals can be, in some level, screw production profile and pushing process specific. However, such potential problem appears to be interesting for deeper analyze by experts from production manager personnel.

Finding Association rules with Statistica

The same experiment like in 5.3 was performed in Statistica 10 Data Miner. This commercial software includes modules for neural network, clusterization, classification trees etc. (StatSoft, 2011) For the association rules mining, module called basket analysis is available. It includes only one rules finding algorithm, the same like used in previous experiment, i.e. Apriori. Rules found by Statistica, shown in Fig. 14, are the same like in previous experiment, where non-commercial Weka software was used.

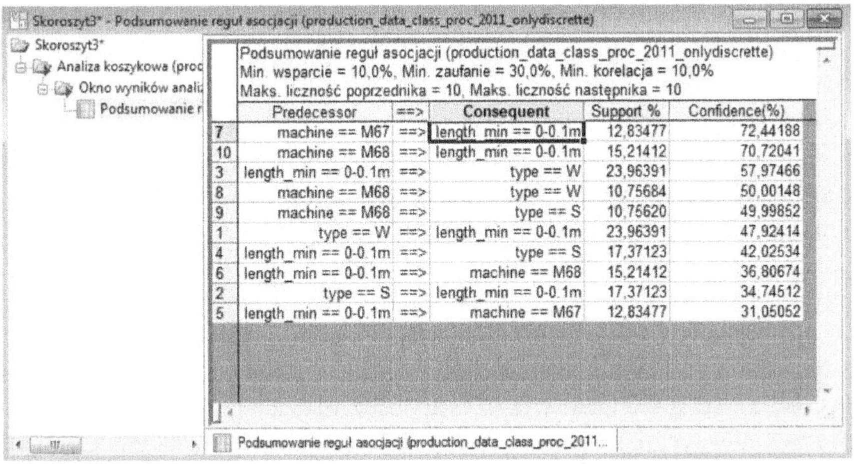

	Predecessor	==>	Consequent	Support %	Confidence(%)
7	machine == M67	==>	length_min == 0-0.1m	12,83477	72,44188
10	machine == M68	==>	length_min == 0-0.1m	15,21412	70,72041
3	length_min == 0-0.1m	==>	type == W	23,96391	57,97466
8	machine == M68	==>	type == W	10,75684	50,00148
9	machine == M68	==>	type == S	10,75620	49,99852
1	type == W	==>	length_min == 0-0.1m	23,96391	47,92414
4	length_min == 0-0.1m	==>	type == S	17,37123	42,02534
6	length_min == 0-0.1m	==>	machine == M68	15,21412	36,80674
2	type == S	==>	length_min == 0-0.1m	17,37123	34,74512
5	length_min == 0-0.1m	==>	machine == M67	12,83477	31,05052

Figure. 14: Association rules found by Statistica's basket analysis module

CONCLUSION

So far the project devoted to preparing platform for industrial implementation of Intelligent Manufacturing System, the hardware and software has been selected and tested in a real production process using the preliminary testbeds. Currently, the platform is mainly used for data collection concerning the production processes, machine operation and operators' work. It also provides

HSI for machines' operators and system end-users, i.e. factory management board. The hardware and software layer, which has already been installed in the factory, has created the need and the basis for the employment of advanced data mining and artificial intelligence techniques and multi-agent software structure in the system. Such techniques could be used for detecting operators' improper actions which could have an influence on machines operation, e.g. increasing downtime duration and number of breakdowns. The analysis of the rules could probably give an answer to the following questions, e.g. what factors and in what manner influence production processes, under what circumstances problems occur, how operators react to diagnostic messages, etc. So far, experiment of finding association rules from the data gathered by production monitoring system was performed for the fixed range of events concerning 12 machines work. Some potentially interesting rules, which can help the factories management personnel to detect and eliminate bottlenecks in the production processes, were discovered. Process of finding association rules is not currently automated, but thanks to open source data mining software Weka, it can be easily integrated with the existing system infrastructure. Remaining work to be done in this area includes extending events attributes list, by adding for each item in data set product type identifier and machine's operators identifier. Experiments with different types of rules discovering algorithms, i.e. Apriori Predictive and Generalized Sequence Patterns, will be also performed. Eventually, the system should automatically generate rules, which help to detect the possibility of the problem occurrence on the basis of historical data. Additionally it should suggest the best solution to the problem, therefore the probability of stoppages will be reduced. New system structures are being developed in order to simplify the system implementation in companies with various production and data resources. The goal is to achieve easy adaptation to the needs of many production sectors. Some of these structures can decrease the cost of system deployment, as popular devices like personal computers can be used on condition that the factory floor is appropriately adaptable, e.g. if there is no oil mist. However, it should be clearly stated that for industrial implementation of IMS, the structure with separate PAC installed for each machine seems to be the most promising. PAC computational power allows running separate machine agents for each machine, so one of new organizational structure, e.g. holonic, can be implemented. Tests of new system versions, especially those with multi-agent solutions, are going to be performed with the usage of the laboratory Flexible Manufacturing System (FMS) testbed, presented in Fig. 15. The FMS testbed consists of an integrated CNC milling machine, robot and vision system. Current results of the project give promising perspectives for an advanced Intelligent Manufacturing System development and implementation with PACs as hardware platform in a real factory for next project stages.

Figure. 15: Laboratory FMS testbed

ACKNOWLEDGMENT

FMS testbed was bought as a part of the project No POPW.01.03.00-18-012/09 from the Structural Funds, The Development of Eastern Poland Operational Programme co-financed by the European Union, the European Regional Development Fund.

REFERENCES

1. Agrawal R., Srikant R. (1994). Fast Algorithms for Mining Association Rules in Large Databases. VLDB'94, Proceedings of 20th International Conference on Very Large Data Bases, pp. 487-499, ISBN 1-55860-153-8, Morgan Kaufmann, Santiago de Chile,Chile, 1994.

2. Agrawal R., Srikant R. (1996). Mining quantitative association rules in large relational tables. Proceedings of the 1996 ACM SIGMOD international conference on Management of data, pp. 1 - 12, ISBN 0-89791-794-4, ACM New York, NY, USA, 1996.

3. Christo C., Cardeira C. (2007). Trends in Intelligent Manufacturing, Proceedings of IEEE International Symposium on Industrial Electronics, pp. 3209-3214, ISBN 978-1-4244-0754-5, Vigo, Spain, June 4-7, 2007.

4. Colombo A., Schoop R., Neubert R. (2006). An Agent-Based Intelligent Control Platform for Industrial Holonic Manufacturing Systems, IEEE Trans. Ind. Elect., vol. 53, no. 1, pp.322-337, ISSN: 0278-0046, Seligenstadt, Germany, February 2006.

5. GlassFish Community (2011). GlassFish Server Open Source Edition, 23.11.2011, Available from: http://glassfish.java.net/.

6. Gong C. (2009). Human-Machine Interface: Design Principles of Visual Information. HumanMachine Interface Design. In: Proc IEEE Conference on Intelligent Human-Machine Systems and Cybernetics, pp 262–265, San Antonio Texas, USA, 2009.

7. Granados F. (2006). Analysis: Industrial Ethernet - Driving the growth, Computing & Control Engineering Journal, vol.17, no.6, pp.14-15, Dec.-Jan. 2006.

8. Hall M., Frank E., Holmes G., Pfahringer B., Reutemann P., Witten I. (2009). The WEKA Data Mining Software: An Update, pp.10-18, SIGKDD Explorations, Volume 11, Issue 1, ACM New York, NY, USA, 2009.

9. Leitão P. (2008). Agent-based distributed manufacturing control: A state-of-the-art survey. Engineering Applications of Artificial Intelligence, No. 22 (7), pp. 979-991, ISBN 0952-1976, ELSEVIER, 2008.

10. Microsoft Developer Network (2011). C# Language Specification, 23.11.2011, Available from http://msdn.microsoft.com/en-us/library/aa645596(v=vs.71).aspx.

11. Mączka T., Czech T. (2010). Manufacturing Control and Monitoring System – Concept and Implementation. Proceedings of IEEE International Symposium on Industrial Electronics, t.1, p.3900-3905, Bari, Italy, July 4-7 2010.

12. Mączka T., Żabiński T. (2011). System for remote machines and operators monitoring -selected elements (in Polish). Pomiary Automatyka Robotyka, p. 62-65, No. 3/2011.

13. Mączka T., Czech T., Żabiński T (2010). Innovative production control and monitoring system as element of factory of future (in Polish). Pomiary Automatyka Robotyka,p.22-25, No. 2/2010.

14. Oborski P. (2004). Man-machine interactions in advanced manufacturing systems. The International Journal of Advanced Manufacturing Technology, Vol. 23. No. 3-4, pp 227- 232, ISSN: 1433-3015, Springer-Link, 2004.

15. Oztemel E. (2010). Intelligent manufacturing systems. L. Benyoucef, B. Grabot, (Ed.) Artificial Intelligence Techniques for Networked Manufacturing Enterprises Management, pp. 1-41,ISBN 978-1-84996-118-9, Springer-Verlag, London, 2010.

16. PostgreSQL (2011). About, 23.11.2011, Available from: http://www.

postgresql.org/about/. ROBO (2011). Student Automation and Robotics scientific circle ROBO (in Polish), 23.11.2011,Available from http://www.robo.kia.prz.edu.pl/.

17. StatSoft (2011). STATISTICA Product Overview, 23.11.2011, Available from http://www.statsoft.com/products/.

18. Żabiński T., Mączka T. (2011). Implementation of Human-System Interface for Manufacturing Organizations. Human-Computer Systems Interaction. Backgrounds and Applications 2, Advances in Soft Computing, Springer-Verlag Co., 2011.

19. Żabiński T., Mączka T., Jędrzejec B. (2009). Control and Monitoring System for Intelligent Manufacturing – Hardware and Communication Software Structure. Proceedings of Computer Methods and Systems, p. 135-140, Kraków, Poland, November 26-27 2009.

Chapter 8

HYBRID MANUFACTURING SYSTEM DESIGN AND DEVELOPMENT

Jacquelyn K. S. Nagel[1] and Frank W. Liou[2]

[1] James Madison University USA
[2] Missouri University of Science and Technology USA

INTRODUCTION

Reliable and economical fabrication of metallic parts with complicated geometries is of considerable interest for the aerospace, medical, automotive, tooling, and consumer products industries. In an effort to shorten the time-to-market, decrease the manufacturing process chain, and cut production costs of products produced by these industries, research has focused on the integration of multiple unit manufacturing processes into one machine. The end goal is to reduce production space, time, and manpower requirements. Integrated systems are increasingly being recognized as a means to meet these goals. Many factors are accelerating the push toward integrated systems. These include the need for reduced equipment and process cost, shorter processing times, reduced inspection time, and reduced handling. On the other hand, integrated systems require a higher level of synthesis than does a single process. Therefore, development of integrated processes will generally be more complex than that of individual unit manufacturing processes, but it could provide simplified, lower-cost manufacturing. Integrated systems in this research area have the ability to produce parts directly from a CAD representation, fabricate complex internal geometries, and form novel material combinations not otherwise possible with traditional subtractive processes. Laser metal deposition (LMD) is an important class of additive manufacturing processes as it provides the necessary functionality and flexibility to produce a wide range of metallic parts (Hopkinson et al., 2006; Liou & Kinsella 2009; Venuvinod & Ma, 2004). Current commercial systems that rely on LMD to produce tooling inserts, prototype parts, and end products are limited by a standard range of material options, building space, and a required postprocessing phase to obtain the desired surface finish and tolerance. To address the needs of industry and expand the applications of a metal deposition process, a hybrid manufacturing system that combines LMD with the subtractive process of machining was developed achieving a fully integrated manufacturing system. Our research into hybrid manufacturing system design and development has lead to the integration of additive and subtractive processes within a single machine footprint such that both processes are leveraged during fabrication. The laser aided manufacturing process (LAMP) system provides a rapid prototyping and rapid manufacturing infrastructure for research and education. The LAMP system creates fully dense, metallic parts and provides all the advantages of the commercial LMD systems. Capabilities beyond complex geometries and good surface finish include: (1) functional gradient material metallic parts where different materials are added from one layer to the next or even from one section to another, (2) seamlessly embedded sensors, (3) part repair to reduce scrap and extend product service life, and (4) thin-walled parts due to the extremely low processing forces (Hopkinson et al., 2006; Liou et al., 2007; Ren et al., 2008). This hybrid system is a very competitive and economical approach to fabricating metallic structures. Hybrid manufacturing systems facilitate a sustainable and intelligent production model and offer flexibility of infrastructure to adapt with emergent technology, customization, and changing market needs (Westkämper, 2007). Consequently, the design strategies, system architecture, and knowledge required to construct hybrid manufacturing systems are vaguely described or are not mentioned at all in literature. The goal of this chapter is to summarize the key research findings related to the design, development, and integration of a hybrid manufacturing process that utilizes LMD to produce fully dense, finished metallic parts. Automation, integration, and control strategies along with the associated issues and solutions are presented as design guidelines to provide future designers with the insight needed to successfully construct a hybrid system. Following an engineering design perspective, the functional and process knowledge of the hybrid system design is explored before physical components are involved. Key results are the system architecture, qualitative modeling, and quantitative modeling and simulation of a hybrid manufacturing process. In summary, this chapter provides an interdisciplinary approach to the design and development of a hybrid manufacturing system to produce metal parts that are not only functional, but also processed to the final desired surface-finished and tolerance. The approach and strategies utilized in this research coalesce to facilitate the design of a sustainable and intelligent production system that offers infrastructure flexibility adaptable with emergent technologies and customizable to changing market needs. Furthermore, the approach to hybrid system design and development can assist in general with integrated manufacturing systems. Applying the strategies to design a new system or retrofit older equipment can lead to increased productivity and system capability.

RELATED WORK

Any process that results in a solid physical part produced directly from a 3D CAD model can be labeled a rapid prototyping process (Kalpakjian & Schmid, 2003; Venuvinod & Ma, 2004). Equally, a process that converts raw materials, layer-by-layer into a product is a rapid prototyping process, but is typically referred to as additive manufacturing or layered manufacturing. Subtractive manufacturing is the process of incrementally removing raw material until the desired dimensions are met. Where additive processes start from the ground up, subtractive processes start from the top down. The combination of manufacturing processes from different processing categories establishes a hybrid manufacturing system. Herein, a hybrid manufacturing system refers to a manufacturing system that is comprised of an additive and subtractive manufacturing process. Both additive and subtractive manufacturing cover a wide range of fabrication processes. For example, additive manufacturing can involve powder-based (e.g., selective laser sintering), liquid-based (e.g., stereolithography) or solid-based (e.g., fused deposition modeling) processes, each using a wide range of materials (Gebhardt, 2003; Kai & Fai, 1997; Venuvinod & Ma, 2004). While traditional subtractive manufacturing is typically reserved for metals, advanced or non-conventional subtractive processes have emerged to handle a greater variety of materials which include electric discharge machining, water jet cutting, electrochemical machining and laser cutting (Kalpakjian & Schmid, 2003). The physical integration of additive and subtractive manufacturing processes, such as laser metal deposition and machining, is the key to leveraging the advantages of each process. The vast domains of additive and subtractive manufacturing have provoked many to test boundaries and try a new concept, in an attempt to discover the next best system that will play a key role in advancing manufacturing technologies. Academic and industry researchers alike have been developing novel, hybrid manufacturing systems, however, the design and integration strategies were not published. On the other hand, a few approaches taken to develop reliable hybrid systems that deliver consistent results, with the majority based on consolidation processes, have published a modest guide to their system design. In following paragraphs, a number of hybrid manufacturing systems are reviewed to give an idea of what has been successful. Beam-directed technologies, such as laser cladding, are very easy to integrate with other processes. Most have been integrated with computer numerically controlled

(CNC) milling machines by simply mounting the cladding head to the z-axis of the milling machine. Kerschbaumer and Ernst retrofitted a Röders RFM 600 DS 5-axis milling machine with an Nd:YAG laser cladding head and powder feeding unit, which are all controlled by extended CNC-control (Kerschbaumer & Ernst, 2004). Similarly, a Direct Laser Deposition (DLD) process utilizing an Nd:YAG laser, coaxial powder nozzle and digitizing system as described by (Nowotny et al., 2003) was integrated into a 3-axis Fadal milling machine. Laser-Based Additive Manufacturing (LBAM) as researched at Southern Methodist University, is a technique that combines an Nd:YAG laser and powder feeder with a custom built motion system that is outfitted with an infrared imaging system (Hu et al., 2002). This process yields high precision metallic parts with consistent process quality. These four systems perform all deposition steps first, and then machine the part to the desired finish, consistent with conventional additive fabrication. Two powder-based manufacturing processes that exhibit excellent material usage and in most cases produced components do not require finishing are Direct Metal Laser Sintering (DMLS) and Laser Consolidation (LC). Using layered manufacturing technology, a DMLS system such as the EOS EOSINT M270 Xtended system, can achieve an acceptable component finish using a fine 20 micron thick metal powder material evenly spread over the build area in 20 micron thick layers (3axis, 2010). Laser Consolidation developed by NRC Canada is a net-shape process that may not require tooling or secondary processing (except interfaces) (Xue, 2006, 2008). Parts produced using these processes exhibit netshape dimensional accuracy and surface finish as well as excellent part strength and material properties. Non-conventional additive processes demonstrate advanced features, alternate additive and subtractive steps, filling shell casts, etc. A hybrid RP process proposed by (Hur et al., 2002) combines a 6-axis machining center with any type of additive process that is machinable, a sheet reverse module, and an advanced process planning software package. What differentiates this process is how the software decomposes the CAD model into machining and deposition feature segments, which maximize the CNC milling machine advantages, and significantly reduces build time while increasing shape accuracy. Laser welding, another hybrid approach, involves a wire feeder, CO2 laser, 5-axis milling center, and a custom PC-NC based control unit that has been used to produce molds for injection molding (Choi et al., 2001). Hybrid-Layered Manufacturing (HLM) as researched by (Akula & Karunakaran, 2006) integrates a TransPulse Synergic MIG/MAG welding process with a conventional milling machine to produce near-net shape tools and dies. This is direct rapid tooling. Welding and face milling operations are alternated to achieve desired layer height and to produce very accurate, dense metal parts. A comparable process was developed at Fraunhofer IPT named

Controlled Metal Build-up (CMB), in which, after each deposited layer the surface is milled smooth (Kloche, 2002). However, CMB utilizes a laser integrated into a conventional milling machine. Song and Park have developed a hybrid deposition process, named 3D welding and milling because a wire-based gas metal arc welding (GMAW) apparatus has been integrated with a CNC machining center (Song & Park, 2006). This process uses gas metal arc welding to deposit faster and more economically. Uniquely, 3D welding and milling can deposit two materials simultaneously with two welding guns or fill deposited shells quickly by pouring molten metal into them. The mold Shape Deposition Manufacturing (SDM) system at Stanford also uses multiple materials to deposit a finished part, however, for a different purpose (Cooper, 1999). A substrate is placed in the CNC mill and sturdy material such as UV-curable resin or wax is deposited to form the walls of a mold, which then is filled with an easily dissolvable material. The top of the mold is deposited over the dissolvable material to finish the mold; once the mold has cooled down the dissolvable material is removed, and replaced with the desired part material. Finally, the sturdy mold is removed to reveal the final part, which can be machined if necessary. Contrary to the typical design sequence (Jeng & Lin, 2001) constructed their own motion and control system for a Selective Laser Cladding (SLC) system and integrated the milling head, which evens out the deposition surface after every two layers. Clearly, each system has its advantages and contributes differently to the RM industry. Although using a CNC milling machine for a motion system is the most common approach to constructing a hybrid system, a robot arm can easily be substituted. This is the case with SDM created at Stanford University (Fessler et al., 1999). The robot arm was fitted with an Nd:YAG laser cladding head which can be positioned accurately, allowing for selective depositing of the material and greatly reducing machining time. Integration of a handling robot can reduce positioning errors and time between operations if the additive and subtractive processes are not physically integrated. Most of the aforementioned systems have been built with versatility in mind and could be set-up to utilize multiple materials or adapted to perform another operation. However, an innovative hybrid system that has very specific operations and capabilities is the variable lamination manufacturing (VLM-ST) and multi-functional hotwire cutting (MHC) system (Yang et al., 2005). The VLM-ST system specializes in large sized objects, up to 3 ft. x 5 ft., by converting polystyrene foam blocks into 3D objects utilizing the turntable of the 4-axis MHC system during cutting; if the object is bigger still, multiple pieces are cut and put together. The design strategy behind several of the reviewed hybrid systems was not emphasized and documented. Thus, key pieces of information for the design and development of hybrid systems are missing which prevents researchers and designers from easily designing and constructing a hybrid system of their own. The information contained within this chapter aims to provide a comprehensive overview of the design, development, and integration of a hybrid manufacturing system such that others can use as a guideline for creating a hybrid system that meets their unique needs.

RESEARCH APPROACH

As previously mentioned, the design strategies, system architecture, and knowledge required to construct a hybrid manufacturing system is vaguely described if mentioned at all in the literature. Consequently, our research approach is mainly empirical. Although our approach relies heavily on observation and experimental data, it has allowed us to identify opportunities for applying theory through modeling and simulation. A major challenge to hybrid manufacturing system design is accurately controlling the physical dimension and material properties of the fabricated part. Therefore, understanding the interaction of all process parameters is key. Layout of the preliminary system architecture provides a basis for qualitative modeling. Independent and dependent process parameters are identified through qualitative modeling, which defines the parameters that require a quantitative understanding for accurate control of the process output. Qualitative models of the hybrid manufacturing process are developed and analyzed to understand both process and functional integration within the hybrid system. This allowed lost, competing or redundant system functionality to be identified and used to inform design decisions. Modeling how the material and information flows through the hybrid system facilitates the development of the automation, integration, and control strategies. Quantitative modeling and simulation of our hybrid manufacturing system concentrates on process control and process planning. Process control modeling is used to predict the layer thickness via an empirical model based on the direct 3D layer deposition, the particle concentration of the powder flow, the nozzle geometry, the carrier gas settings, and the powder-laser interaction effects on the melt pool. Process planning models are used to automate part orientation, building direction, and the tool path. These models assist with resolving the challenges of the laser deposition process including building overhang structures, producing precision surfaces, and making parts with complex structures. Revisiting the preliminary system architecture design with the knowledge gained from qualitative and quantitative modeling has resulted in a system architecture that enables accurate and efficient fabrication of 3D structures. Decomposition of the system architecture allows for direct mapping of customer needs and requirements to the overall system architecture.

HYBRID MANUFACTURING SYSTEM

The laser aided manufacturing process (LAMP) lab at Missouri University of Science and Technology (formerly University of Missouri-Rolla) houses a 5-axis hybrid manufacturing system, which was established by Dr. Liou and other faculty in the late 1990s. This system entails additive-subtractive integration, as shown in Fig. 1, to build a rapid prototyping/ manufacturing infrastructure for research and education at Missouri S&T. Integration of this kind was planned specifically to gain sturdy thin wall structures, good surface finish, and complex internal features, which are not possible by a LMD or machining system alone. Overall, the system design provides greater build capability, better accuracy, and better surface finish of structures with minimal post-processing while supporting automated control. Applications of the system include repairing damaged parts (Liou et al., 2007), creating functionally gradient materials, fabrication of overhang parts without support structures, and embedding sensors, and cooling channels into specialty parts.

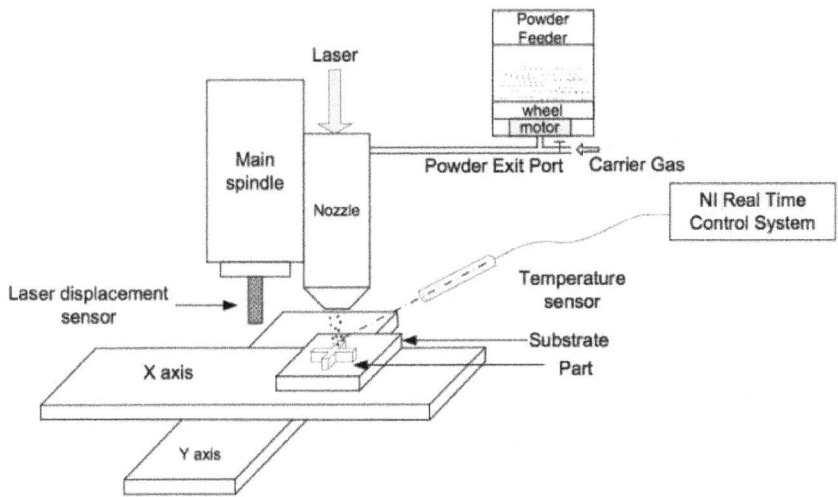

Figure. 1: Five-axis Hybrid Manufacturing Process (Adapted from Tang et al., 2007)

The LAMP hybrid system is comprised of five subsystems or integration elements: process planning, control system, motion system, manufacturing process, and a finishing system. Equipment associated with subsystem is described in the following paragraphs and summarized in Table 1. The LAMP process planning system is a in-house layered manufacturing or slicing software that imports STL models from a commercial CAD package to generate a description that specifies melt pool length (mm), melt pool peak temperature, clad height (mm) and sequences of operations. The objective of the process planning software is to integrate the five-axis motion and deposition-machining hybrid processes. The results consist of the subpart information and the build/ machining sequence (Ren et al., 2010; Ruan et al., 2005). To generate an accurate machine tool path a part skeleton, which calculates distance and offset edges or boundaries, is created of the CAD model. Distance, gradient, and tracing functions were modified to allow more complicated and unconnected known environments for successful implementation with the LAMP hybrid manufacturing system. Basic planning steps involve determining the base face, extracting the skeleton, decomposing a part into subparts, determining build sequence and direction for subparts, checking the feasibility of the build sequence and direction for the machining process, and optimization of the deposition and machining.

Table 1: LAMP Hybrid System Equipment

Hybrid Manufacturing Subsystems	LAMP Hybrid System Equipment
Process Planning	Commercial and in-house CAD software
Motion	Fadal 3016L 5-axis VMC
Manufacturing Process	Nuvonyx 1kW diode laser, Bay State Surface Technologies 1200 powder feeder
Control	NI RT PXI chassis & LabVIEW, Mikron temperature sensor, Omron laser displacement sensor, Fastcom machine vision system
Finishing	Fadal 3016L 5-axis VMC

True 3D additive manufacturing processes can be achieved with a 5-axis machining center without additional support structures (Ruan et al., 2005), as opposed to 2.5D that is afforded by a 3-axis machine. Therefore the motion subsystem for the LAMP hybrid manufacturing system is a 5-axis Fadal 3016L VMC, which also constitutes the finishing subsystem. Servo motors control the motion along the axes as compared with crank wheels and shafts in conventional machine tools. The Fadal VMC is controlled via G and M codes either entered at the control panel or remotely fed through an RS-232 connection. The main manufacturing process of the hybrid system is laser metal deposition, the additive manufacturing process. Metal powder is melted using a 1kW diode laser while the motion system traverses in response to the tool path generated by the process planning software, thereby creating molten tracks in a layer-by-layer fashion on a metal substrate. Layers are deposited with a minimum thickness of 10μm. The melt pool temperature is between 1000°C and 1800°C, depending on the material (e.g. H13 tool steel, Titanium alloy), but is less than 2000°C. A commercial powder delivery system, designed for plasma-spraying processes carries the steel or titanium powder to the substrate via argon. The cladding head is mounted to the z-axis of the Fadal VMC to fully utilize the motion system and provide the opportunity to machine the fabricated part at any point in the deposition process by applying a translation algorithm. The beam focusing optics, beam splitter for out-coupling the process radiation from the laser beam path, water cooling connections, powder feeder connections, and various sensors (optional) are located within the cladding head. Built in to the cladding head are pathways for metal powder to travel through to the laser beam path in a concentric form, therefore, releasing metal powder in a uniform volume and rate. Quartz glass is used to focus the laser beam and water carried from the chiller to the cladding head by small plastic hoses reduces the wear on the focusing optics. Overall, the LMD subsystem includes equipment for lasing, cooling, and powder material delivery. Control of the hybrid manufacturing subsystems require a versatile industrial controller and a range of sensors to acquire feedback. The National Instruments Real Time Control System (NI RT System) provides analog and digital I/O ports and channels, DAC, RS-232, and ADC for controlling all the subsystems of the hybrid system. The control system contains a PXI 8170 Processor, 8211 Ethernet card, 8422 RS-232 card, 6527 Digital I/O card, 6711 Analog Output card, 6602 Timing I/O card, 6040E Multi-function card, and an SCXI Controller with 1304 card. PCI eXtensions for Instrumentation (PXI) is a PC-based platform for measurement and automation systems. PXI combines PCI electrical-bus features with the modular, Eurocard packaging of Compact PCI, and then adds specialized synchronization buses and key software features. Signal Conditioning Extension for Instrumentation (SCXI) is a front-end signal conditioning and switching system for various measurement devices, including plug-in data acquisition devices. Our control system offers modularity, expandability, and high bandwidth in a single, unified platform. System feedback is acquired through temperature and laser displacement sensors. An Omron Z4M-W100 laser displacement sensor is used to digitally determine the cladding head height above the substrate. There are danger zones and safe zones that the nozzle can be with respect to the substrate. Output of the displacement sensor is -4 to +4 VDC which is converted into a minimum and maximum distance value, respectively. The temperature sensor is a Mikron MI-GA5-LO non- contact, fiber-optic, infrared temperature sensor. It was installed onto the Z-axis of the VMC with a custom, adjustable fixture. The set-up for data acquisition of the melt pool temperature, while deposition takes place is at an angle of 42°, 180 mm from the melt pool and sampling every 2 ms. There is also a machine vision system, a Fastcom iMVS-155 CMOS image sensor, to watch the melt pool in real-time. It has also been used to monitor melt pool geometry and assist with our empirical approach to fine tune process parameters.

HYBRID MANUFACTURING SYSTEM DESIGN AND DEVELOPMENT

The critical success factors of an integrated system are quality, adaptability, productivity and flexibility (Garelle & Stark, 1988). Inclusion of additive fabrication technology in a traditional subtractive manufacturing system inherently addresses these four factors. Nevertheless, considering the four success factors during the initial design phase will ensure that the resultant manufacturing system will meet short and long term expectations, be reliable, and mitigate system obsolescence. In order for hybrid manufacturing systems to become a widespread option they must also be an economical solution. Dorf and Kusiakpoint out that the three flows within a manufacturing system i.e. material, information, and cost, which "should work effectively in close cooperation for efficient and economical manufacturing" (Dorf & Kusiak, 1994). This section reviews the qualitative and quantitative modeling efforts of material and information as well as the system architecture design that incorporates the knowledge gained through modeling. Cost modeling for the hybrid system has only been temporal, however, a cost benefit analysis as proposed in (Nagel & Liou, 2010) could be performed to quantify the savings.

System Architecture

Initially, the LAMP system design was integrated only through the physical combination of the laser metal deposition process (additive manufacturing) and the machining center (subtractive manufacturing). Also, each subsystem housed a separate controller, including the LMD and VMC, which required manual control of the hybrid system. Reconfiguring the LAMP hybrid system to utilize a central control system, increased communication between the subsystems and eliminated the need for multiple people. Moreover, the process can be controlled and monitored from a remote location, increasing the safety of the manufacturing process. The hybrid manufacturing system architecture follows the modular, integration element structure as defined in (Nagel & Liou, 2010). Figure 2 shows the direct mapping of customer needs and requirements to the overall system architecture as well as the dependency relationships. Build geometry, surface finish, and material properties are the needs relating directly to the finished product. Efficient operation and flexibility are the system requirements to be competitive and relate directly to the system itself.

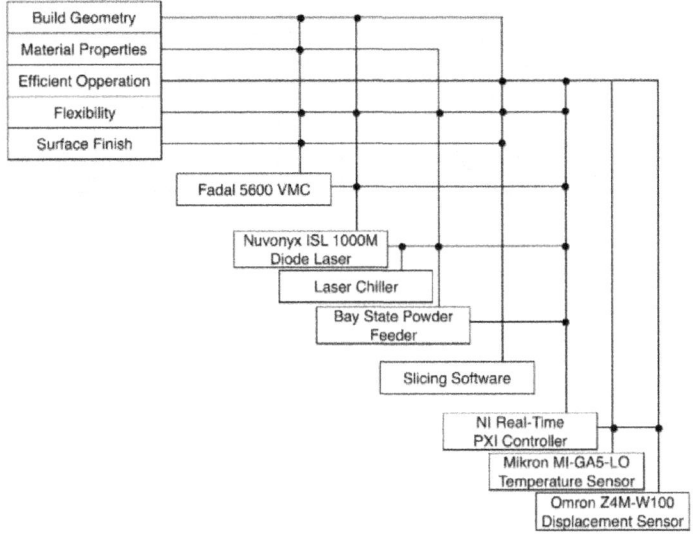

Figure. 2: LAMP Hybrid System Architecture

Qualitative Modeling

Qualitative modeling efforts are focused on understanding process parameters and the flow of the process. Modeling the process parameter interactions uncovers the independent and dependent process parameters where as modeling the manufacturing process identifies opportunities for optimization. The following subsections summarize how qualitative modeling has been used to gain knowledge of the relationships among process parameters and resources utilized in each step of the hybrid manufacturing process.

Independent Process Parameters

The major independent process parameters for the hybrid manufacturing system include the following: laser beam power, process speed, powder feed rate, incident laser beam diameter, and laser beam path width (path overlap) as shown in Fig. 3 (Liou et al., 2001). Other parameters such as cladding head to surface distance (standoff distance), carrier gas flow rate, absorptivity, and depth of focus with respect to the substrate also play important roles.

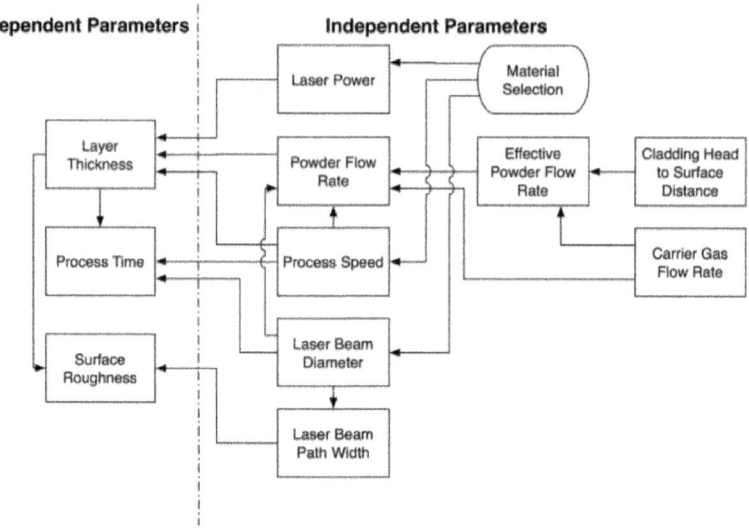

Figure. 3: LAMP Hybrid System Process Parameters (Adapted from Liou et al., 2001)

The layer thickness process parameter is directly related to the power density of the laser beam and is a function of incident beam power and beam diameter. Generally, for a constant beam diameter, the layer thickness increases with increasing beam power provided corresponding powder feed rate. It was also observed that the deposition rate increased with increasing laser power (Weerasinghe & Steen, 1983). Powder mass flow rate is another important process parameter which directly affects layer thickness. However, effective powder flow rate, which includes powder efficiency during the LMD process, turned out to be a more important parameter (Lin & Steen, 1998; Mazumder et al., 1999). Also the factor that most significantly affected the percent powder utilization was laser power. The cladding head nozzle is set up to give a concentric supply of powder to the melt pool, and due to the nature of the set-up, the powder flow is hour glass-shaped. The powder flow initially is unfocused as it passes through the cladding head, but the nozzle guides the powder concentrically towards its center, and essentially "focuses" the beam of powder. The smallest diameter focus of the powder "beam" is dependent upon the design of the cladding head nozzle. Also, if the laser beam diameter becomes too small as compared to the powder beam diameter, e.g., 100μm, much of the supplied powder will not reach the melt pool. Thus, there will be unacceptably low powder utilization. Process speed has a big impact on the process output. In general, decreasing process speed increases the layer thickness. There is a threshold to reduce process speed, however, as too much specific energy (as defined in Section 5.2.2) will cause tempering or secondary hardening of previous layers (Mazumder et al., 1997). Process speed should be well chosen since it has strong influence on microstructure.

The laser beam diameter parameter is one of the most important variables because it determines the power density. It can be difficult to accurately measure high power laser beams. This is partly due to the shape of the effective beam diameter (e.g., Gaussian, Top hat) and partly due to the definition of what is to be measured. Single isotherm contouring techniques such as charring paper and drilling acrylic or metal plates are well known but suffer from the fact that the particular isotherm they plot is both power and exposure time dependent. Multiple isotherm contouring techniques overcome these difficulties but are tedious to interpret. Beam path width or beam width overlap has a strong influence on surface roughness. As the deposition pass overlap increases, the valley between passes is raised due to the overlap therefore reducing the surface roughness. Powder that has adhered to the surface, but has not melted will be processed in successive passes. In order to obtain the best surface quality, the percent pass overlap should be increased as much as possible. Conversely, to decrease the surface roughness, the deposition layers should be kept as thin as possible.

Dependent Process Parameters

The major dependent process parameters of the hybrid manufacturing system are: layer thickness, surface roughness, and process time (Fig. 3). Other dependent parameters such as hardness, microstructure, and mechanical properties should also be considered, but in this chapter we will focus only on the parameters related with physical dimension. There is a large range of layer thicknesses as well as deposition rates that can be achieved using LMD. However, part quality consideration puts a limit on optimal deposition speeds. Both the layer thickness and the volume deposition rates are affected predominately by the specific energy and powder mass flow rate. Here, specific energy (SE) is defined as: $SE = p/(Dv)$, where p is the laser beam power, D is the laser beam diameter and v is the process speed. Also it has been well known that actual laser power absorbed in the melt pool is not the same as the nominal laser power measured from a laser power monitor due to reflectivity and other plasma related factors depending on the materials (Duley, 1983). The use of adjusted specific energy is thus preferable. Considering the factors, there is a positive linear relationship between the layer thickness and adjusted specific energy for a range of powder mass flow rates (Liou et al., 2001). Surface roughness was found to be highly dependent on the direction of measurements with respect to the deposited metal (Liou et al., 2001; Mazumder et al., 1999). In checking the surface roughness, at least four directions should be tested from each sample; the length and width direction on the top surface, and the horizontal and vertical directions on thin walls. Since the largest roughness on each sample is of primary interest, measurements should be only taken perpendicular to the deposition direction on the top surface and in the vertical direction on the walls, based on our experiments. The overall deposition processing time is mainly dependent upon the layer thickness per slice, process speed, and laser beam diameter. The processing conditions need to be optimized prior to optimizing the processing time, since the processing time is directly influenced by the processing conditions. If the laser beam diameter is increased, the specific energy and power density will be decreased under the same process condition, that means, a lower deposition rate unless the laser power and powder mass flow rate are increased correspondingly. Similarly, when the process speed is increased the independent process parameters should be optimized accordingly.

Process Modeling

Process modeling used to model the hybrid manufacturing system aims to optimize the sequence with which the material flows through the system (Shunk, 1992; Wang, 1997). Following the process modeling approach by (Nagel et al., 2009), process events and tasks within each event were identified. Part A of Fig. 4 shows the manually controlled hybrid manufacturing process. Decomposition of the system process aided with identification of integration points to reduce the number of steps and events within the process resulting in significant time savings. Part B of Fig. 4 shows the optimized process.

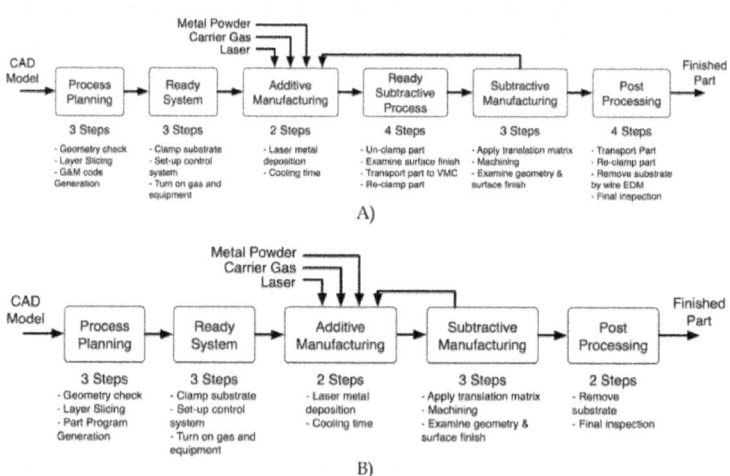

Figure. 4: LAMP Hybrid System Process Models, A) Before integration and optimization, B) After integration and optimization

Once the process is clearly laid out, the motion system and control system can be accurately defined. Reconfiguring the LAMP hybrid system elements to utilize a central control system increased communication between the subsystems and eliminated the need for multiple people. Moreover, the process can be controlled and monitored from a remote location, increasing the safety of the manufacturing process. Supplementary improvements were made to the process planning software, laser metal deposition subsystems, and the VMC. In efforts to eliminate the separate VMC computer, required only to upload machine code via direct numerical control, the RS-232 communication protocol utilized by Fadal was reverse engineered and implemented via LABVIEW. The laser, cooling, and powder material delivery subsystems of the laser metal deposition process are equipped with external control ports, but were not utilized in previous system configurations. Subsequently, all subsystems and modules were directly connected to the control system hardware so external control could be utilized. Initializing communications among the LAMP subsystems became the foundation for the control system software. Off-line, the in-house layered manufacturing software only converted CAD models into layer-by-layer slices of machine code to create the tool path. With the central control system now in place, the in-house layered manufacturing software was changed to generate machine code, laser power, and powder flow commands, which together comprise a part program and are distributed via the control system software. Overall, manufacturing process integration has resulted in modularity, easy maintenance, and process improvement. Thus, increasing system productivity and capability.

Quantitative Modeling and Simulation

Quantitative modeling and simulation provides a theoretical foundation for explaining the phenomena observed through empirical research. Additionally, detailed modeling assists with developing a quantitative understanding of the relationship between independent process parameters and dependent process parameters. Understanding the relationships among parameters affords accurate control of physical dimension and material properties of the part. While separate modeling efforts were undertaken, outputs of one model feed into another. The following subsections summarize how quantitative modeling has been used to develop a theoretical understanding of the LAMP hybrid manufacturing process.

Melt Pool Modeling and Simulation

Melt pool geometry and thermal behavior control are essential in obtaining consistent building performances, such as geometrical accuracy, microstructure, and residual stress. A 3D model was developed to predict the thermal behavior and geometry of the melt pool in the laser material interaction process (Han et al., 2005). The evolution of the melt pool and effects of the process parameters were investigated through modeling and simulations with stationary and moving laser beam cases. When the intense laser beam irradiates on the substrate surface, the melt pool will appear beneath the laser beam and it moves along with the motion of the laser beam. In order to interpret the interaction mechanisms between laser beam and substrate the model considers the melt pool and adjacent region. The governing equations for the conservation of mass, momentum and energy can be expressed in following form:

$$\frac{\partial}{\partial t}(\rho) + \nabla \cdot (\rho \mathbf{V}) = 0 \tag{1}$$

$$\frac{\partial}{\partial t}(\rho \mathbf{V}) + \nabla \cdot (\rho \mathbf{V}\mathbf{V}) = \nabla \cdot (\mu_l \frac{\rho}{\rho_l} \nabla \mathbf{V}) - \nabla p - \frac{\mu_l}{K} \frac{\rho}{\rho_l}(\mathbf{V} - \mathbf{V}_s) + \rho g \tag{2}$$

$$\frac{\partial}{\partial t}(\rho h) + \nabla \cdot (\rho \mathbf{V} h) = \nabla \cdot (k\nabla T) - \nabla \cdot (\rho (h_l - h)(\mathbf{V} - \mathbf{V}_s)) \tag{3}$$

where ρ, V, p, μ, T, k, and h are density, velocity vector, pressure, molten fluid dynamic viscosity, temperature, conductivity, and enthalpy, respectively. K is the permeability of mushy zone, Vs is moving velocity of substrate with respect to laser beam and subscripts of s and l represent solid and liquid phases. Since the solid and liquid phases may coexist in the same calculation cell at the mushy zone, mixed types of thermal physical properties are applied in the numerical implementation. The liquid/vapor interface is the most difficult boundary for numerical implementation in this model since many physical phenomena and interfacial forces are involved there. To solve those interfacial forces the level set method is employed to acquire the solution of the melt pool free surface (Han et al., 2005). To avoid numerical instability arising from the physical property jump at the liquid/vapor interface, the Heaviside function $H(\varphi)$ is introduced to define a transition region where the physical properties are mollified.

The energy balance between the input laser energy and heat loss induced by evaporation, convection and radiation determines surface temperature. Laser power, beam spot radius, distance from calculation cell to the beam center, and the absorptivity coefficient are used to calculate the laser heat influx. Heat loss at the liquid/vapor interface is computed in terms of convective heat loss, radiation heat loss and evaporation heat loss. The roles of the convection and surface deformation on the heat dissipation and melt pool geometry are revealed by dimensionless analysis. It was found that interfacial forces including thermo-capillary force, surface tension and recoil vapor pressure considerably affect the melt pool shape and fluid flow. Quantitative comparison of interfacial forces indicates that recoil vapor pressure is dominant under the melt pool center while thermo-capillary force and surface tension are more important at the periphery of the melt pool. For verification, the intelligent vision system was utilized to acquire melt pool images in real time at different laser power levels and process speeds, and the melt pool geometries were measured by cross-sectioning the samples obtained at various process conditions (Han et al., 2005). Simulation predictions were compared to experimental results for both the stationary laser case and moving laser case at various process conditions. Model prediction results strongly correlate to experimental data. An example of melt pool shape comparison between simulation and experiment for the moving laser beam case is shown in Fig. 5.

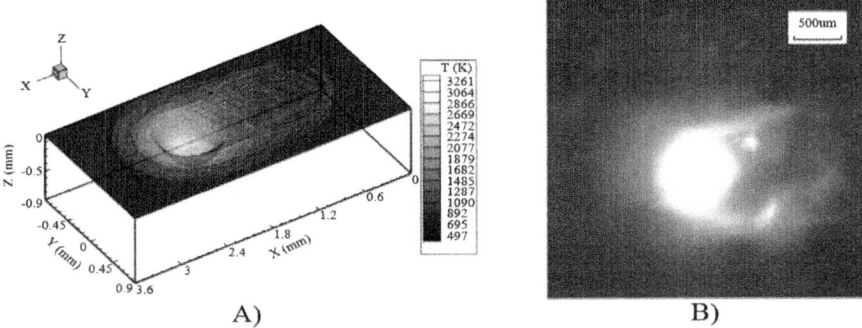

A)   B)

Figure. 5: Melt pool shape comparison, A) Simulation result of melt pool shape and surface temperature, B) Experimental result of melt pool shape (Adapted from Han et al., 2005)

Powder Flow Dynamics Modeling and Simulation

Analysis of metallic powder flow in the feeding system is of particular significance to researchers in order to optimize the LMD fabrication technique. Powder flow simulation holds a critical role in understanding flow phenomena. A stochastic Lagrangian model for simulating the dispersion behavior of metallic powder, or powder flow induced by nonspherical particle-wall interactions, is described (Pan & Liou, 2005). The numerical model also takes into consideration particle shape effects. In wall-bounded, gas-solid flows, the wall collision process plays an important role and is strongly affected by particle shape. Non-spherical effects are considered as the deviation from pure spheres shows induced particle dispersion, which has a great impact on the focusibility of the powder stream at the laser cladding head nozzle exit. The parameters involved in non-spherical collision are analyzed for their influencing factors as well as their interrelations.

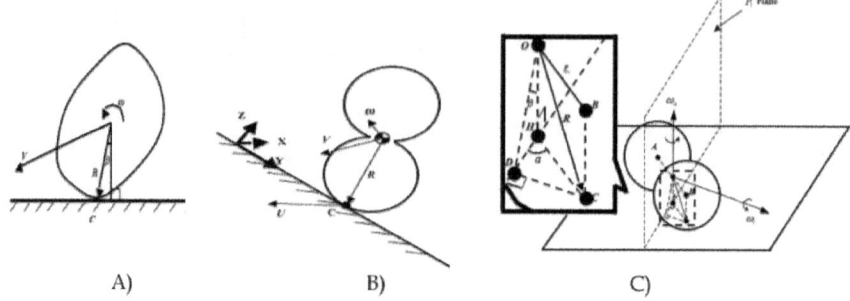

A)   B)   C)

Figure. 6: Particle Collision Diagrams, A) 2D non-spherical particle-wall collision model, B) Local coordinate for collision model, C) of 3-D non-spherical particle-wall collision model (Adapted from Pan & Liou, 2005)

The parameters involved in the 2-D non-spherical model include β and R, as shown in Fig. 6, Part A, where β indicates how much the contact point C deviates from the foot of a vertical from the gravity center of the particle and R shows the actual distance between the contact point and the gravity center. The collision coordinate system used to describe the 3D collision dynamics is defined in Fig. 6, Part B. The contact velocity is computed from:

$$U = V + \omega \times R \tag{4}$$

where V is the particle translational velocity vector, ω is the angular velocity vector, and R is the vector connecting particle mass center to contact point C. The change in the contact point velocity can be obtained by the following equation:

$$\Delta U = [\frac{1}{m} I - R^x J^{-1} R^x] \Delta P \tag{5}$$

where m is particle mass, I is the 3x3 identity matrix, and R^x is the canonical 3x3 skewsymmetric matrix corresponding to R, ΔP denotes the impulse delivered to the particle in the collision, and J^{-1} is the inverted inertia tensor in the local coordinate. As shown in Fig. 6,

Part C, a cluster that consists of two identical spheres with equal radius r represents the nonspherical powder particle of the 3D model. This representation leads to generalized modeling of satelliting metallic powder particles. Wall roughness also effects powder dispersion behavior, therefore in this model the roughness effect was included by using the model and parameters proposed by (Sommerfeld & Huber, 1999). The instantaneous impact angle is assumed to be composed of the particle trajectory angle with respect to the plane wall and a random component sampled from a Gaussian distribution function. It was also assumed that each collision has 30% possibility to be non-spherical, which implies the stochastic model was applied in 30% of the total collisions during the feeding process simulation. Simulations using the spherical model (0% non-sphericity) were also conducted. The non-spherical model successfully predicts the actual powder concentration profile along the radial and axial directions, whereas the spherical particle model underestimates the dispersion and results in a narrow spread of the stream along the radial direction. When compared to the experimental results, the 3D simulated powder stream is in strong agreement, which demonstrates validation of the model. The model also predicts the peak powder concentration or focal point of the power stream for specific cladding head nozzle geometry. It is essential to establish a well-focused powder stream at the exit of the nozzle and to know the ideal stand-off distance in order to increase powder catchment in the melt pool, achieve high material integrity, and reduce material waste.

Tool Path Modeling and Simulation

Process planning, simulation, and tool path generation allows the designer to visualize and simulate part fabrication prior to manufacturing to ensure a successful process. Adaptive multi-axis slicing, collision detection, and adaptive tool path pattern generation for LMD as well as tool path generation for surface machining are the key advantages to the integrated process planning software developed for the LAMP hybrid system (Ren et al., 2010). Basic planning steps involve determining the base face and extracting the skeleton of an input CAD model (Fig. 7, top left). The skeleton is found using the centroidal axis extraction algorithm (Fig. 7, top right). Based on the centroidal axis, the part is decomposed into subcomponents and for each sub-component a different slicing direction is defined according to build direction. In order to build some of the components, not only translation but also rotation will be needed to finish building the whole part because different subcomponents have different building directions (Fig. 7, bottom middle), and the laser nozzle direction is always along the z-axis. After the decomposition (Fig. 7, bottom left) results are obtained, the relationship among all the components is determined, and a building relationship graph is created. From the slicing results and build directions, collision detection is determined. Collision detection is implemented by Boolean operation, which is an intersection operation, on a simulation (Ren et al., 2010). If the intersection result of the updated CAD model and the cladding head nozzle is not empty, then collision will happen in the real deposition process. The deformation of the CAD model following the building relationship graph includes two categories: positional deformation and dimensional deformation. Positional change means translation or rotation of the CAD model. The dimensions will change after every slicing layer is finished. For every updated model, collision needs to be checked before the next slicing layer is added. Following the collision detection algorithm, if a potential collision is detected the sequence of the slicing layers is reorganized (Ren et al., 2010). The output of the collision detection algorithm will be the final list of slicing layers, which comprise the actual building sequence when manufacturing the part.

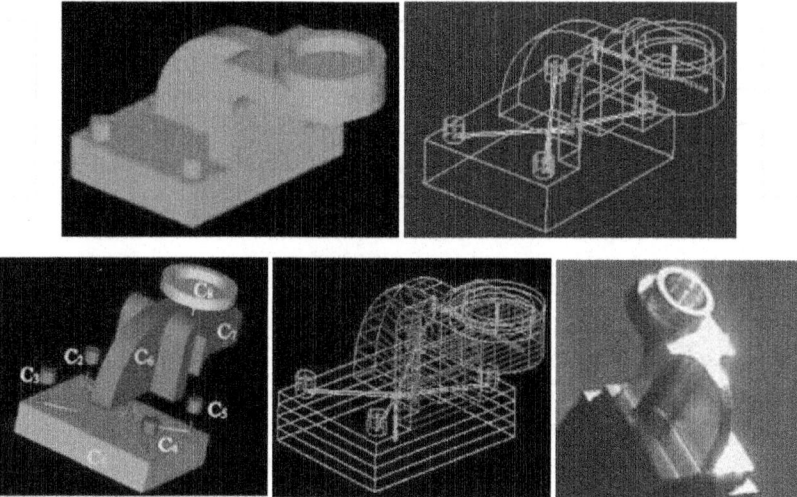

Figure. 7: Process Planning and Fabrication of 3D Part (Adapted from Liou et al., 2007; Ren et al. 2010)

The final piece of process planning is tool path generation. Common tool path patterns are the raster, contour-parallel offsetting, zig-zag, and interlaced. Each pattern has advantages and disadvantages. The adaptive deposition tool path algorithm considers each pattern when predicting the possibility of deposition voids. The goals of the algorithm are to adjust the tool path to remove deposition voids and increase time efficiency. Multiple tool path patterns may be used during fabrication and the algorithm may also prescribe alternating the appropriate tool path pattern when necessary. Surface finish machining is a sequential step used after deposition to improve manufacturing quality after deposition is finished. The process planning software allows the designer to specify the machining parameters including the feed rate, spindle speed, and depth of cut before determining the number of machining cycles necessary. As with LMD, alignment will be also integrated for 3D geometries to achieve the accuracy without reloading the deposited part to be machined. Again, the tool path will be generated such that a collision-free machining tool path will be generated for the deposited part. A visibility map algorithm (Ruan & Liou, 2003) is applied to detect the collision between the tool and the deposited part.

The final process planning step is to generate the part program. This step is the bridge between the algorithmic results of process planning, quantitative modeling of process parameters, and the realistic operational procedures as well as parameters of the 5-axis manufacturing environment. It will build the map of the process planning results and the real operational parameters and then interpret the final planning tool path as the corresponding movements of the hybrid manufacturing system. The software will combine and refine those movements and translate them into machine executable code. Resulting in a text file composed of three columns of data to needed for the control system to command the laser, powder feeder, and motion system (Ren et al., 2010). The final set of operations is based on the building relationship graph, build directions that avoid collisions, tool path, and time required.

HYBRID MANUFACTURING SYSTEM INTEGRATION

During the course of this research several integrated manufacturing system designs were analyzed to identify what characteristics comprise a successful hybrid system. Based on this background research, and the experiences of working with and refining the LAMP system, the key elements of a hybrid manufacturing system were identified. The five key elements represent an effective way to design a hybrid manufacturing system, as compared to a reconfigurable or mechatronic design, because the identified elements contain necessary subsystems, are easily modularized, and advocate the use of off-the-shelf hardware and software. Within an integrated system, each element acts as a separate subsystem affording a stable modular design (Gerelle & Stark, 1988). A strategy for controlling the integrated LMD and machining processes, the 5-axis motion system, and the data corresponding streams provides the basis for fully automating the system. Considering scalability, our integration strategy emphasizes modularity of the integrated components but also modularity of the controlling software. Our control strategy allows data streams to be easily added or removed. Furthermore, our design allows an operator to optimize the control strategy for a particular geometry.

Physical Integration

Obstacles arise during the development of any manufacturing system; however, by identifying obstacles and solutions the industry as a whole can benefit. Outside of cost and yield, the obstacles of developing a hybrid manufacturing system discussed here cover a range of topics. Table 2 summarizes the obstacles associated with the physical integration of the LAMP hybrid system and provides documented solutions.

Table 2: Physical Integration Issues and Solutions

Issue	Solution	Result	Reference
Adding the laser cladding head to a VMC	A platen with precisely tapped holes for the cladding head mounted to the Z-axis of the VMC	Laser cladding head is securely mounted and future equipment or fixtures can be added	
Protection of Equipment	Retract laser head or position it far enough away from the machining head	Protect laser nozzle	Kerschbaumer & Ernst, 2004
	Mount a displacement sensor on the Z-axis	When cladding head gets too close to X-Y axes the process halts	
Unknown communication protocol	Use reverse engineering to figure out communication protocol	Subsystems can be controlled from a central control system	Stroble et al., 2006
Quality control	Implement control charts, pareto charts, etc.	Manual quality control	Starr, 2004
	Sensor feedback utilized by closed-loop controllers	Automated quality control	Boddu et al., 2003; Doumanidis & Kwak, 2001; Hu & Kovacevic, 2003; Tang, 2007
Transition between additive and subtractive processes	Apply a translation matrix that repositions the X-Y axis for the desired process	Accurate positioning for machining or LMD	
Placement of sensors to monitor melt pool due to high heat of the LMD process	Mount the sensitive vision system in-line with the laser using a dichromatic mirror attachment for the cladding head, and custom hardware mounted to the platen holds the temperature probe at an acceptable viewing angle	Sensors are safe, and the LMD process is accessible	Boddu et al., 2003; Tang & Landers, 2010

The documented information in Table 2 does not address every possible integration obstacle, but is meant to be comprehensive from what is found in the literature and personal experience. Issues outside of integration, such as material properties can be found in (Nagel & Liou, 2010). After central control, integration, and modularity were enforced in the LAMP hybrid system, manufacturing defects and time were significantly reduced, and safety was significantly increased. Material integrity was improved as the laser could be precisely commanded on/off or pulsed as needed during deposition. Furthermore, by integrating the laser power and powder flow commands into the process planning software and automating the distribution of commands, functionally graded parts were manufactured effortlessly.

Software Integration

Utilization of a central control system directly resulted in automation of the LAMP hybrid system and allowed unconventional possibilities to be explored. To achieve the central controller, a framework consisting of a multi-phase plan and implementation methodology was developed. The automation framework involves controlling the laser, powder feeder, and motion system, and utilizing sensor feedback, all through the NI PXI control subsystem. Open and closed-loop controllers were designed, along with compatibility and proper module communication checking.

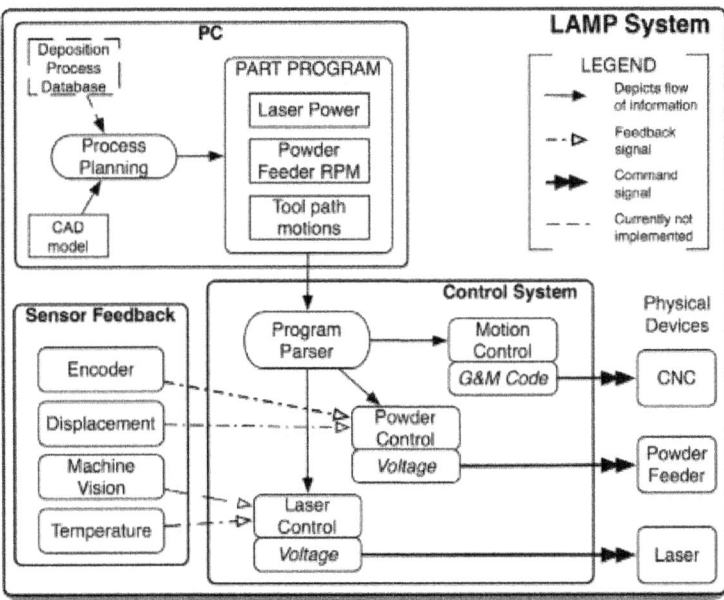

Figure. 8: LAMP Hybrid System Communication Schematic

Moreover, compensation for undesired system dynamics, delays and noise were considered to ensure a reliable and accurate automated manufacturing process. The result of the automation framework is an automated deposition program (developed in LabVIEW) with a customized graphical user interface and data recording capabilities. Figure 8 is a visual description of the LAMP hybrid system communications layout, including process planning that occurs outside the control system. Once process planning completes the part program, with laser power and powder mass flow rate commands in the form of voltages, the control system parses through the information to automatically fabricate the desired part. While commands are being sent to the physical devices, sensors are monitoring the process and sending feedback to the control system simultaneously, allowing parameters to change in real-time. Unique to the LAMP hybrid system is that the hardware and software are both modular. The automated deposition program that is executed by the control system has three different modes: dry-run, open-loop control, and closed-loop control. Fundamental code within the automated deposition program is shared amongst each of the modes, much like the control system is central to the LAMP hybrid system. Additional portions of code that control the laser, control the powder feeder, utilize feedback, or simply read in, display and record data from sensors are turned on or off by each mode. Code modularity prevents large amounts of the control system software from being rewritten when equipment is upgraded or subsystems are replaced. During dry-run mode only machine code is distributed by the control system, allowing the user to monitor the VMC motions without wasting materials and energy. This mode is primarily utilized to check uncertain tool paths for instances when the laser should be shut off or when a tool path transition seems too risky. For instance, transitions from one geometry to another may rotate longer than desired at one point causing a mound to form and solidify, which destroys the overall part geometry and could collide with the laser cladding nozzle. Open-loop and closed-loop control modes are provided for fabricating parts and include system monitoring and data acquisition features. The modular software allows multiple closed-loop controllers optimized for a particular geometry to be added as research is completed, such as a feed forward controller (Tang et al., 2007) that regulates powder flow to the melt pool for circular, thin walled structures or thin walled structures with many arcs.

CONCLUSION

In an effort to shorten the time-to-market, decrease the manufacturing process chain and cut production costs, research has focused on the integration of multiple manufacturing processes into one machine; meaning less production space, time, and manpower needed. An integrated or hybrid system has all the same features and advantages of rapid prototyping systems, plus provides a new set of features and benefits. Moreover, hybrid manufacturing systems are increasingly being recognized as a means to produce parts in material combinations not otherwise possible and have the ability to fabricate complex internal geometries, which is beyond anything that can be accomplished with subtractive technologies alone. Internal geometries such as complex conformal cooling channels provide better product thermal performance, which additive fabrication processes create them with ease, giving the manufacturer a better product with little extra cost. As manufacturers and customers dream up more complex products, requiring more advanced equipment and software, hybrid systems will emerge. In short, integrating additive and subtractive technologies to create new manufacturing systems and processes is going to advance the manufacturing industry in today's competitive market. Modeling and simulation, both qualitative and quantitative, were shown to be an integral part of hybrid system design and development as well as motivate areas of research that a pure empirical approach does not reveal. Although this research is focused on integrating additive and subtractive processes, the general principles can also be applicable to integrating other unit manufacturing processes (NRC, 1995). Integrated processes can combine multiple processes that fall within the same family, such as different material removal processes, or they can combine processes that are in different unit process families, such as a mass-change process and a microstructure-change process. The results can lead to significant processing breakthroughs for low-cost, high-quality production. Future work includes applying integrated process and product analysis to various hybrid processes that integrate different manufacturing processes and applying the hybrid system concept to other types of configurations, such as those that include robots. Model-based simulation reveals various new opportunities for simultaneous improvement of part quality, energy and material efficiencies, and environmental cleanness. Thereby, accelerating the hybrid integration process. Other work includes applying an open architecture for the hybrid controller, as such an architecture avoids the difficulties of using proprietary technology and offers an efficient environment for operation and programming, ease of integrating various system configurations, and provides the ability to communicate more effectively with CAD/CAM systems and factory-wide information management systems.

ACKNOWLEDGMENT

This research was supported by the National Science Foundation Grant # IIP-0822739, the U.S. Air Force Research Laboratory, and the Missouri S&T Intelligent Systems Center. We would also like to thank all the Missouri S&T LAMP lab researchers that have contributed over the years to make this body of work possible, especially Zhiqiang Fan, Lijun Han, Heng Pan, Lan Ren, Jianzhong Ruan, and Todd Sparks. Their support is greatly appreciated.

REFERENCES

1. Akula, S. & Karunakaran, K.P. (2006). Hybrid Adaptive Layer Manufacturing: An Intelligent Art of Direct Metal Rapid Tooling Process. Robotics and Computer-Integrated Manufacturing, Vol. 22, No. 2, pp. 113-123, 0736-5845

2. Boddu, M.R., Landers, R.G., Musti, S., Agarwal, S., Ruan, J. & Liou, F. W. (2003) System Integration and Real-Time Control Architecture of a Laser Aided Manufacturing Process. Proceedings of SFF Symposium, 1053-2153, Austin, TX, August, 2003

3. Choi, D.-S., Lee, S.H., Shin, B.S., Whang, K.H., Song, Y.A., Park, S.H. & Lee, H.S. (2001) Development of a Direct Metal Freeform Fabrication Technique Using Co2 Laser Welding and Milling Technology. Journal of Materials Processing Technology, Vol.113, No. 1-3, (June 2001), pp. (273-279), 0924-0136.

4. Cooper, A.G. (1999). Fabrication of Ceramic Components Using Mold Shape Deposition Manufacturing. Doctor of Philosophy Thesis, Stanford, USA.

5. Dorf, R.C. & Kusiak, A. (1994). Handbook of Design, Manufacturing and Automation, J. Wiley and Sons, 0471552186, New York, N.Y.

6. Doumanidis, C. & Kwak, Y.-M. (2001) Geometry Modeling and Control by Infrared and Laser Sensing in Thermal Manufacturing with Material Deposition. Journal of Manufacturing Science and Engineering, Vol. 123, No. 1, pp. 45-52, 1087-1357

7. Duley, W.W. (2003). Laser Processing and Analysis of Materials, Plenum Press, 0306410672, New York

8. Fessler, J.R., Merz, R., Nickel, A.H. & Prinz, F.B. (1999) Laser Deposition of Metals for Shape Deposition Manufacturing. Proceedings of SFF Symposium, 1053-2153, Austin, TX, August, 1999

9. Gebhardt, A. (2003). Rapid Prototyping, Hanser Publications, 9781569902813, Munich Gerelle, E.G.R. & Stark, J. (1988). Integrated Manufacturing: Strategy, Planning, and Implementation, McGraw-Hill, 0070232350, New York

10. Han, L., Liou, F. & Musti, S. (2005). Thermal Behavior and Geometry Model of Melt Pool in Laser Material Process. Journal of Heat Transfer, Vol. 127, No. 9, pp. 1005, 0022-1481

11. Hopkinson, N., Hague, R.J.M. & Dickens, P.M. (2006) Rapid Manufacturing: An Industrial Revolution for the Digital Age, John Wiley, 0470016132, Chichester, England Hu, D. & Kovacevic, R. (2003).

Sensing, Modeling and Control for Laser-Based Additive Manufacturing. International Journal of Machine Tools & Manufacture, Vol. 43, No. 1,(January 2003), pp. 51-60, 0890-6955

12. Hu, D., Mei, H. & Kovacevic, R. (2002). Improving Solid Freeform Fabrication by LaserBased Additive Manufacturing. Proceedings of the Institution of Mechanical Engineers, Part B: Journal of Engineering Manufacture, Vol. 216, No. 9, pp. 1253-1264, 1253-1264

13. Hur, J., Lee, K., Hu, Z. & Kim, J. (2002). Hybrid Rapid Prototyping System Using Machining and Deposition. Computer-Aided Design, Vol. 34, No.10, (September 2002), pp. 741-754, 0010-4485

14. Jeng, J.-Y. & Lin, M.-C. (2001). Mold fabrication and modification using hybrid processes of selective laser cladding and milling. Journal of Materials Processing Technology, Vol.110, No. 1, (March 2001), pp 98-103, 0924-0136

15. Kai, C.C., & Fai, L.K. (1997). Rapid Prototyping: Principles & Applications in Manufacturing, John Wiley, 9810245165, New York.

16. Kerschbaumer, M., & Ernst, G. (2004). Hybrid Manufacturing Process for Rapid High Performance Tooling Combining High Speed Milling and Laser Cladding. Proc.

17. 23rd International Congress on Applications of Lasers & Electro-Optics (ICALEO), 0912035773 , San Francisco, CA, October 2004

18. Klocke, F. (2002). Rapid Manufacture of Metal Components. Fraunhofer Institute for Production Technology, IPT

19. Kalpakjian, S., & Schmid, S.R. (2003). Manufacturing Processes for Engineering Materials. Pearson Education, Inc., 9780130453730, Upper Saddle River, N.J

20. Lin, J. & Steen, W.M. (1998). Design characteristics and development of a nozzle for coaxial laser cladding. Journal of Laser Applications, Vol. 10, No. 2, pp. 55-63, 1042-346X

21. Liou, F.W., & Kinsella, M. (2009). A Rapid Manufacturing Process for High Performance Precision Metal Parts. Proceedings of SME Rapid 2009 Conference and Exhibition, Paper No. TP09PUB18, Schaumburg, IL, May, 2009

22. Liou, F., Slattery, K., Kinsella, M., Newkirk, J., Chou, H.-N., Landers, R. (2007). Applications of a Hybrid Manufacturing Process for Fabrication and Repair of Metallic Structures, Rapid Prototyping Journal, Vol. 13, No. 4, pp. 236–244, 1355-2546

23. Liou, F.W., Choi, J., Landers, R.G., Janardhan, V., Balakrishnan, S.N.,

& Agarwal, S. (2001). Research and Development of a Hybrid Rapid Manufacturing process. Proceedings of SFF Symposium, 1053-2153, Austin, TX, August, 2001

24. Mazumder, J., Choi, J., Nagarathnam, K., Koch, J. & Hetzner, D. (1997). The Direct Metal Deposition of H13 Tool Steel for 3-D Components. JOM, Vol. 49, No. 5, pp.55-60 1047-4838

25. Mazumder, J., Schifferer, A., & Choi, J. (1999). Direct Materials Deposition: Designed Macro and Microstructure", Materials Research Innovations, Vol. 3, No. 3, (October 1999), pp.118-131, 1432-8917

26. NRC (National Research Council) (1995). Unit Manufacturing Processes: Issues and Opportunities in Research, The National Academies Press

27. Nagel, J.K.S. & Liou, F. (2010). Designing a Modular Rapid Manufacturing Process. Journal of Manufacturing Science and Engineering, Vol. 132, No. 6, (December 2010), pp. 061006,1087-1357

28. Nagel, R., Hutcheson, R., Stone, R., & Mcadams, D., (2009). Process and Event Modeling for Conceptual Design. Journal of Engineering Design, Vol. 22, No. 3, (March 2011),pp.145-164, 0954-4828

29. Nowotny, S., Scharek, S., & Naumann, T. (2003). Integrated Machine Tool for Laser Beam Cladding and Freeforming. Proc. Thermal Spray 2003: Advancing the Science &Applying the Technology, 9780871707857, Orlando, FL, May, 2003

30. Pan, H. & Liou, F. (2005). Numerical simulation of metallic powder flow in a coaxial nozzle for the laser aided deposition process. Journal of Materials Processing Technology, Vol. 168, No. 2, pp. 230-244, 0924-0136

31. Ren, L., Sparks, T., Ruan, J. & Liou, F. (2010). Integrated Process Planning for a Multiaxis Hybrid Manufacturing System. Journal of Manufacturing Science and Engineering,Vol. 132, No. 2, pp. 021006, 1087-1357

32. Ren, L., Padathu, A.P., Ruan, J., Sparks, T., & Liou, F.W. (2008). Three Dimensional Die Repair Using a Hybrid Manufacturing System. Proceedings of SFF Symposium, 1053-2153, Austin, TX, August, 2008

33. Ruan, J., Eismas-Ard, K., & Liou, F. W., 2005, "Automatic Process Planning and Toolpath Generation of a Multiaxis Hybrid Manufacturing System," Journal of Manufacturing Processes, 7(1), pp. 57-68.

34. Ruan, J., & Liou, F.W. (2003) Automatic Toolpath Generation for Multi-axis Surface Machining in a Hybrid Manufacturing System. Proc. of ASME IDETC/CIE,0791837009, Chicago, IL., September, 2003

35. Shunk, D.L. (1992). Integrated Process Design and Development,

Business One Irwin, 9781556235566, Homewood, IL.

36. Sommerfeld, M., & Huber, N. (1999). Experimental analysis and modeling of particle-wall collisions. International of multiphase flow, Vol. 25, pp.1457-1489, 0301-9322

37. Song, Y.-A., & Park, S. (2006). Experimental Investigations into Rapid Prototyping of Composites by Novel Hybrid Deposition Process. Journal of Materials Processing Technology, Vol. 171, No. 1, (January 2006), pp. 35-40, 0924-0136

38. Starr, M. K., 2004, Production and Operations Management, Atomic Dog, 1592600921, Cincinnati, OH.

39. Stroble, J.K., Landers, R.G., & Liou, F.W. (2006). Automation of a Hybrid Manufacturing System through Tight Integration of Software and Sensor Feedback. Proceedings of SFF Symposium, 1053-2153, Austin, TX, August, 2006

40. Tang, L. & Landers, R.G. (2010). Melt Pool Temperature Control for Laser Metal Deposition Processes, Part I: Online Temperature Control," Journal of Manufacturing Science and Engineering, Vol. 132, No. 1, (February 2010), pp. 011010, 1087-1357

41. Tang, L., Ruan, J., Landers, R., & Liou, F. (2007). Variable Powder Flow Rate Control in Laser Metal Deposition Processes, Proceedings of SFF Symposium, 1053-2153, Austin, TX,August, 2007

42. Venuvinod, P.K., & Ma, W. (2004). Rapid Prototyping Laser-Based and Other Technologies, Kluwer Academic Publishers, ISBN: 978-1-402-07577-3, Boston

43. Wang, B. (1997). Integrated Product, Process and Enterprise Design, Chapman & Hall,0412620200, New York.

44. Weerasinghe V.W. & Steen, W.M. (1983). Laser Cladding with Pneumatic Powder Delivery. Proceedings of Laser Materials Processing, Los Angeles, CA, January 1983

45. Westkämper, E. (2007). Digital Manufacturing In The Global Era, In: Digital Enterprise Technology, Pedro Cunha and Paul Maropoulos, pp. 3-14, Springer, 978-0-387-49864-5, New York

46. Xue, L. (2006). Laser Consolidation – a One-Step Manufacturing Process for Making NetShaped Functional Aerospace Components. Proc. SAE International Aerospace Manufacturing and Automated Fastening Conference & Exhibition, Toulouse, France,September, 2006

47. Xue, L. (2008). Laser Consolidation--a Rapid Manufacturing Process for Making Net-Shape Functional Components, Industrial Materials

Institute, National Research Council of Canada, London, Ont.

48. Yang, D.Y., Kim, H.C., Lee, S.H., Ahn, D.G. & Park, S.K. (2005). Rapid Fabrication of LargeSized Solid Shape Using Variable Lamination Manufacturing and Multi-Functional Hotwire Cutting System. Proceedings of SFF Symposium, 1053-2153, Austin, TX,August, 2005

49. 3axis, Direct Metal Laser Sintering (Dmls-Eos), Retrieved on Jan. 14, 2010, http://www.3axis.us/direct_metal_laser_slintering_dmls.asp

Chapter 9

DIGITAL MANUFACTURING SUPPORTING AUTONOMY AND COLLABORATION OF MANUFACTURING SYSTEMS

Hasse Nylund and Paul H Andersson

Tampere University of Technology Finland

INTRODUCTION

This chapter discusses on the challenges and opportunities of digital manufacturing supporting the decision making in autonomous and collaborative actions of manufacturing companies. The motivation is the change towards more networked collaboration caused by, for example, globally distributed markets and specialization of manufacturing companies to their core competences, their autonomous activities. This situation has led to increasingly complex manufacturing activities in the manufacturing network and the importance of collaboration has become a critical factor. In most cases companies seek to respond to the challenges through cooperation rather than expanding their own operations. The autonomy means that the parties involved in the manufacturing activities do their own tasks by themselves independently from other parties while the collaboration involves the activities that one party cannot do by itself and therefore, co-operation of several parties are required. This kind of situation can be clearly seen in networked manufacturing activities involving several companies, but similarly, inside a company and its one facility, same kind of autonomous and collaborative activities can be recognized. In the discussion, the dimensions of autonomy and collaboration are considered in designing and developing manufacturing systems, as well as in improving the daily operations. The rest of this Chapter is structured as follows. Section 2 discusses on the main issues behind the research, including Competitive and Sustainable Manufacturing, changeability in manufacturing as well as support from digital manufacturing. In Section 3, a structure for manufacturing systems and entities is proposed, which is the base for the design and development activities of manufacturing systems discussed in Section 4. An academic research environment is introduced in Section 5 describing several of the theoretical aspects discussed before. Section 6 gives a brief conclusion on the topics discussed.

BACKGROUND

The focus of the discussion is on mechanical engineering industry of discrete part manufacturing for business-to-business (B2B) industry, including their part manufacturing and product assemblies. These kinds of products are typically highly customized and tailored to customer needs and requirements with low or medium demand (Lapinleimu, 2001). This type of production usually involves several companies and is formed as a supply network. For example, the production includes a main company, its suppliers and suppliers of suppliers as well as customers and customers of customers. The current manufacturing paradigm, in the above context, has evolved from the early craft manufacturing via mass manufacturing towards mass customization. Typical characteristics that have been recognized include (Andersson, 2007):

- Globally local systems spread over industrial ecosystems and manufacturing networks of their own pros and cons.

- Managing the networked manufacturing, where the importance of procurement and management of knowledge flow increase.

- Specialization to one's core competences and collaborating with others in the manufacturing network.

Early discussions considered whether these characteristics could be fulfilled with developing existing flexible manufacturing systems (FMSs), or to shift to reconfigurable manufacturing systems (RMSs) paradigm. At some point, more ambitious goals were set with the aim to describe a manufacturing system with autonomous entities having the needed level intelligence to be changeable to organize themselves to altered situations, and to identify what new entities will be required. At the same time, a manufacturing system is required to be competitive in order to survive in the markets as well as sustainable to reduce or eliminate unwanted activities and outputs.

Competitive and Sustainable Manufacturing

The well-known definition of sustainability is: "The Sustainable Development is development that meets the needs of the present without compromising the ability of future generations to meet their own needs" (World Commission on Environment and Development [WCED], 1987), thereafter (WCED, 1987). This political statement is the root cause for today's key global challenges and related problems that call for a drastic change of paradigm from economic to sustainable development. Competitive Sustainable Manufacturing (CSM) is seen as a fundamental enabler of such change (Jovane, 2009). Sustainable development has been recently increasingly emphasized around the world; in Europe (Factories of Future Strategic Roadmap and the Manufuture initiative), the USA (Lean and Mean), and Japan (Monozukuri and New JIT). The CSM paradigm widens the classical view of sustainability to interact with the Social, Technological, Economical, Environmental, and Political (STEEP) context (AdHoc, 2009). Sustainable manufacturing is a multi-level approach where product development, manufacturing systems and processes as well as enterprise and supply chain levels need to be considered, with metrics identified for each level (Jawahir et al., 2009).

The CSM is one of the strategic research areas within the Department of Production Engineering (TTE) at Tampere University of Technology (TUT). Figure 1 presents the main areas of the CSM approach, consisting of three main pillars, Sustainable, Lean and Agile Manufacturing. Lean manufacturing aims to combine the advantages of craft and mass production, while avoiding the drawbacks such as the high costs of craft production and rigidity of mass production systems (Womack et al., 1990). For example, the Lean Enterprise Institute (2008) defines Lean manufacturing as "a business system for organizing and managing product development, operations, suppliers, and customer relations that requires less human effort, less space, less capital, and less time to make products with fewer defects to precise customer desires, compared with the previous system of mass production."

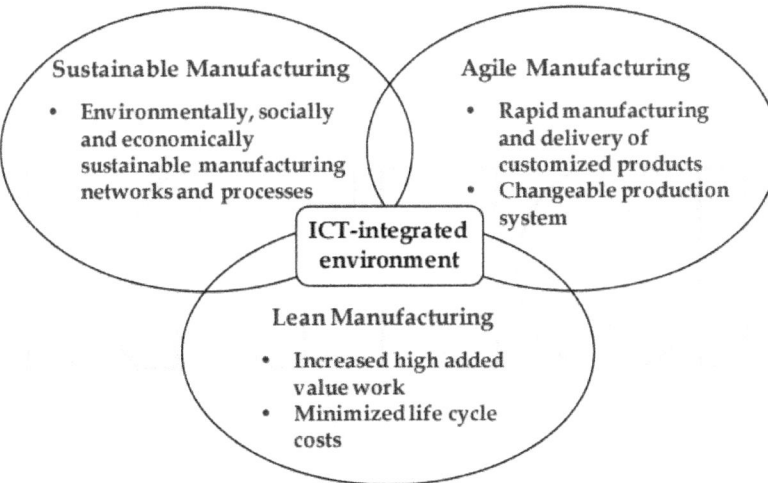

Figure. 1: The cornerstones of the CSM at the Department of Production Engineering (Nylund et al., 2010)

Agile manufacturing can be defined as an enterprise level manufacturing strategy of introducing new products into rapidly changing markets (Nagel & Dove, 1991) and an organizational ability to thrive in a competitive environment characterized by continuous and sometimes unforeseen change (Kidd, 1994). Agile manufacturing highlights the need to adapt to changes in the business environment, and generally agility is defined as ability to react to and take advantage of changes and opportunities, see for example (Sharifi & Zhang, 1999; Gould, 1997). Sustainable development is the development that meets the needs of the present without compromising the ability of future generations to meet their own needs (WCED, 1987). It consists of three structural pillars namely society, environment, and economy, whilst at the same time it also involves operational aspects such as the consumption of resources, natural environment, economic performance, workers, products, social justice and community development (Jayachandran et al., 2006). When these three pillars of Lean, Agile, and Sustainable are considered as one system, Lean emphasized the stability of a system that can be referred as the autonomy while agility adds the needed capability to change to new situations, therefore focusing more on the collaboration. These two have their main focus on economic issues while sustainability adds the viewpoints of energy and environmentally friendly manufacturing.

Changeability in Manufacturing Systems

Wiendahl et al. (2007) suggest changeability as an important factor in the competitiveness of manufacturing companies in addition to the classical factors of cost, quality, and time. Changeability is defined on the five structuring levels of an enterprise: changeover ability, reconfigurability, flexibility, transformability and agility. Agility, which was discussed in the context of CSM, is seen from a manufacturing enterprise level and refers to the ability of an enterprise to effect changes in its systems, structure and organization (Gunasekaran & Yusuf, 2002). Transformability is changeability at a factory level. It includes, for example, facilities, organization and employees. The whole factory is oriented towards the market to offer the right products and services (Wiendahl et al., 2007). Into a detailed level of manufacturing activities the term changeover ability is used. It is related to single workstations that perform manufacturing processes in order to manufacture product features. Reconfigurability and flexibility are the most widely examined structuring levels of changeability in the context of manufacturing systems. An FMS is configured to deal with part variations within its scope. The functionality and capacity of FMSs are pre-designed, while flexibility is inherent and built-in a priori (ElMaraghy, 2005). Because of the fixed flexibility of FMS, it is not flexible enough for rapid and cost-effective reconfiguration in response to changing markets (Mehrabi et al., 2000). An RMS is composed of generalpurpose hardware and software modules that are reused in reconfiguration tasks. Modules are replaced or added only if necessary. An RMS has the ability to change capacity and functionality to bring about the needed flexibility, i.e. to bring about exactly the functionality and capacity needed exactly when needed (Koren, 1999).

Support from Digital Manufacturing

The tools and principles of digital manufacturing, factories, and enterprises can offer significant value to all aspects of manufacturing systems during their life cycles. However, there are no commonly used or agreed definitions for those, but they usually share the idea of managing the typically isolated and separate manufacturing activities as a whole by the means of Information and Communications Technology (ICT) (Nylund and Andersson, 2011). Typical examples often found from the definitions, based on literature, are (see, for example: Bracht & Masurat, 2005; Maropoulos, 2003; Souza et al., 2006):

- An integrated approach to develop and improve product and production engineering technologies.
- Computer-aided tools for planning and analysing real manufacturing systems and processes.
- A collection of new technologies, systems, and methods.

Typical tools and principles of digital manufacturing on different structuring levels are, for example (Kühn, 2006):

- Computer-aided technologies, such as computer-aided design (CAD) and computeraided manufacturing (CAM), e.g. offline programming for virtual tool path generation to detect collisions, analyse material removal and optimise cycle times.
- Visual interaction applications, e.g. virtual environments and 3D-motion simulations that offer realistic 3D graphics and animations to demonstrate different activities.
- Simulation for the reachability and sequences of operations as well as internal work cell layout and material handling design. These include, for example, realistic robotics simulation (RRS) and ergonomics simulation.
- Discrete event simulation (DES) solutions including the need for and the quantity of equipment and personnel as well as evaluation of operational procedures and performance. DES can also be focused on e.g. factories and supply chain or network sales and delivery processes as well as to complex networked manufacturing activities, including

logistical accuracy and delivery reliability of increasing product variety.

The above are examples of typical application areas of digital manufacturing. In each case, the activities rely on up-to-date and accurate information and knowledge. The total information and knowledge of a manufacturing system can be explained with explicit and tacit components (Nonaka and Takeuchi, 1995). The explicit part of the knowledge can be described precisely and presented formally in ICT-systems. The skills of humans are explained as the tacit dimension of knowledge, which, presented digitally, may lead to unclear situations and can be wrongly understood. The importance of the transformation from tacit to explicit knowledge has been recognized as one of the key priorities of knowledge presentation (Chryssolouris et al., 2008). Challenges exist both in the autonomous and collaborative parts of the digitally presented manufacturing entities. The internal part should include only the needed information and knowledge to fully describe the autonomous activities while the collaboration mostly relies on effective sharing of information and knowledge and therefore both the communication language and content should be described formally. Effective knowledge management consists of four essential processes: creation, storage and retrieval, transfer, as well as application, which are dynamic and continuous phenomenon (Alavi and Leidner, 2001). Examples of the application areas of the digital part are:

- Email messages, Internet Relay Chat (IRC), Instant Messaging, message boards and discussion forums.

- More permanent information and knowledge derived from the informal discussions, stored in applications such as Wikipedia.

- Internet search engines and digital, such as dictionaries, databases, as well as electronic books and articles

- Office documents, such as reports, presentations, as well as spreadsheets and database solutions.

- Formally presented information systems, such as Enterprise Resource Planning (ERP), Product Data Management (PDM), and Product Lifecycle Management (PLM).

The importance of the possibilities offered by ICT tools and principles is ever more acknowledged, not only in academia, but also in industry. The Strategic Multi-annual Roadmap, prepared by the Ad-Hoc Industrial Advisory Group for the Factories of the Future Public-Private Partnership (AIAG FoF PPP), lists ICT as one of the key enablers for improving manufacturing systems (AdHoc, 2010). The report describes the role of ICT at three levels; smart, virtual, and digital factories.

- Smart factories involve process automation control, planning, simulation and optimisation technologies, robotics, and tools for competitive and sustainable manufacturing.

- Virtual factories focus on the value creation from global networked operations involving global supply chain management.

- Digital factories aim at a better understanding and the design of manufacturing systems for better product life cycle management involving simulation, modelling and management of knowledge.

Both digitally presented information and knowledge as well as computer tools and principles for modelling, simulation, and analysis offer efficient ways to achieve solutions for design and development activities. General benefits include, for example:

- Experiments in a digital manufacturing system, on a computer model, do not disturb the real manufacturing system, as new policies, operating procedures, methods etc. can be experimented with and evaluated in advance in a virtual environment.

- Solution alternatives and operational rules can be compared within the system constraints. Possible problems can be identified and diagnosed before actions are taken in the real system.

- Modelling and simulation tools offer real-looking 3D models, animations, and visualisations that can be used to demonstrate ideas and plans as well as to train company personnel.

- Being involved in the process of constructing the digital manufacturing system tasks increases individuals' knowledge and understanding of the system. The experts in a manufacturing enterprise acquire a wider outlook compared to their special domain of knowledge as they need to gather information also outside their daily operations and responsibilities.

STRUCTURE OF MANUFACTURING ENTITIES AND SYSTEMS

The proposed structure of manufacturing systems consists of manufacturing entities as well as their related domains and activities. An entity, being autonomous, is something that has a distinct existence and can be differentiated from other entities. The term 'entity' has similarities to other terms, such as: object, module, agent, actor, and unit. A domain is an expert area in which two or more entities are collaborating. Domains have certain roles in the system and their own responsibilities and specific objectives. An activity is a set of actions that accomplish a task that is related to the entities and domains, as well as to their context.

Structure of Manufacturing Entities

Figure 2 illustrates the general viewpoints of the proposed structure of manufacturing entities. The structure is explained with internal structure of individual manufacturing entities. It is derived from the principles behind the term 'holon' and the concept of Holonic Manufacturing Systems (HMS). The term holon comes from the Greek word 'holos', which is a whole and the suffix '–on', meaning a part. Therefore the term holon means something that is at the same time a whole and a part of some greater whole (Koestler, 1989). In HMS, holons are autonomous and co-operative building blocks of a manufacturing system, consisting of information processing part and often a physical processing part (Van Brussel et al., 1998). In this approach, the information part is divided into digital and virtual parts differentiating the digitally presented information and knowledge from the computer models representing the existing or future possible real manufacturing entities. The digital part barely exists as clearly consisting separate part. It can be distributed in several information systems both globally and locally and in information rich computer models, the virtual parts of the manufacturing entities.

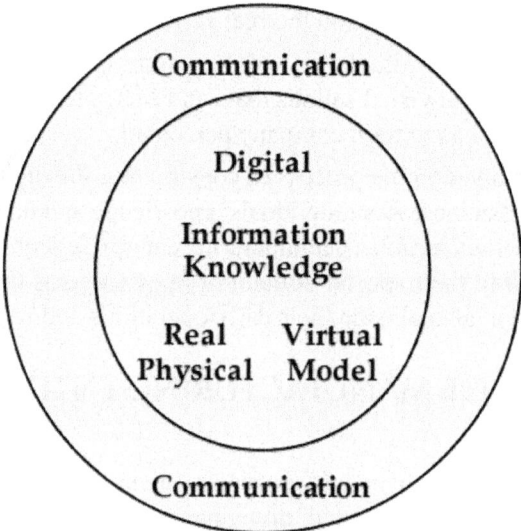

Figure. 2: Internal structure of manufacturing entities

The digital, virtual, and real parts combined present the autonomy of a manufacturing entity. The communication part is responsible of both the language and content of the messages between manufacturing entities. Therefore, it enables the manufacturing entities to collaborate with each other

(Nylund & Andersson, 2011). As the autonomous entities exist distributed, independently from each other, they can be developed separately. At the same time, the communication part enables the investigation of the entities in an integrated fashion, and to develop the whole system they form. The division into digital, virtual, and real is intentionally missing the tacit dimension, as it is intended to be used in decision making processes by humans, based on their skills and knowledge. At the end, the humans are the ones that are making the decisions, or are the ones that are creating the decision making mechanisms.

Structure of Manufacturing Systems

A manufacturing system consists of manufacturing entities with different roles as well as their related domains and activities. Figure 3 shows a general presentation of manufacturing entities of products, orders, and resources as well as their connecting domains of process, production, and business. The focus is on the manufacturing activities that are related to the transformation of raw material to finished products and their associated services as well as the flow of information and knowledge that is related to the physical manufacturing of customer orders.

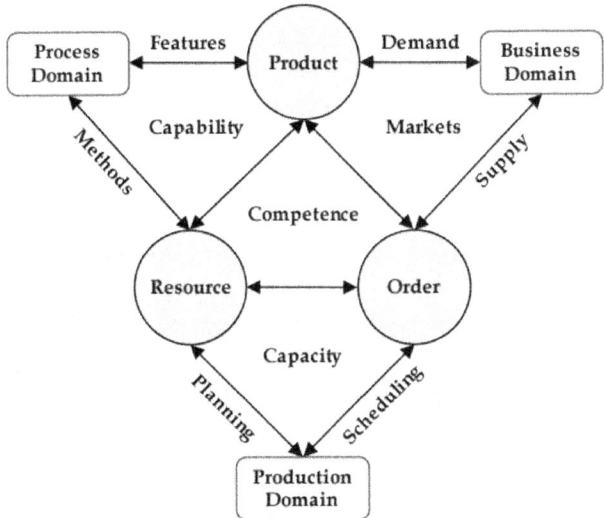

Figure. 3: Structure of manufacturing systems

The proposed structure is loosely based on the HMS reference architecture ProductResource-Order-Staff Architecture (PROSA) (Van Brussel et al., 1998). The PROSA explains the relations between the entities with the information and knowledge they exchange while in this approach the relations are explained with activities occurring between the entities. Brief descriptions of the entities and domains are:

- Products represent what the manufacturing system offers to its customers. The characteristics of the products specify the requirements for the manufacturing system, i.e. what the system should be able to do.

- Resources embody what is available to manufacture the products. The characteristics of the resources determine what kinds of products can be manufactured.

- Orders represent instances of products that are ordered by customers. They define the volume and variation requirements of the products ordered, as well as the capacity and scalability requirements for the manufacturing system.

- The process domain represents the capabilities that are needed to manufacture the products. It connects the development activities of products and resources.

- The production domain defines the capacity and scalability to manufacture changing volumes and variations in customer orders. It handles the material and information flow of the manufacturing system.

- The business domain is responsible for markets, i.e. for the right products being available for the customers to gain enough orders.

Structuring Levels in Manufacturing

A fractal is an independently acting manufacturing entity that can be precisely described (Warnecke, 1993). Fractals are structured bottom-up, building fractals of a higher order.

Entities at the higher levels always assume only those responsibilities in the processes which cannot be fulfilled in lower order (Strauss & Hummel, 1995). This is similar to holons and holarchies, as at every fractal level of holons the level above is the holarchy of the holons at a lower level. Similarly, the autonomy of the holons is not considered in the holarchy, but instead dealing with and organizing the co-operation of the holons is the responsibility of the holarchy. In Figure 4, four different structuring levels, manufacturing units, stages, plants, and networks, are distinguished.Manufacturing units correspond to individual machine tools that have certain manufacturing methods. The units are designated to manufacture the features of work pieces that have similarities in, for example, size and shape as well as tolerances and material properties. Typical areas are computer-aided design (CAD) and computer-aided manufacturing (CAM), e.g. offline programming for virtual tool path generation to detect collisions, analyse material removal and optimise cycle times (Kühn, 2006).

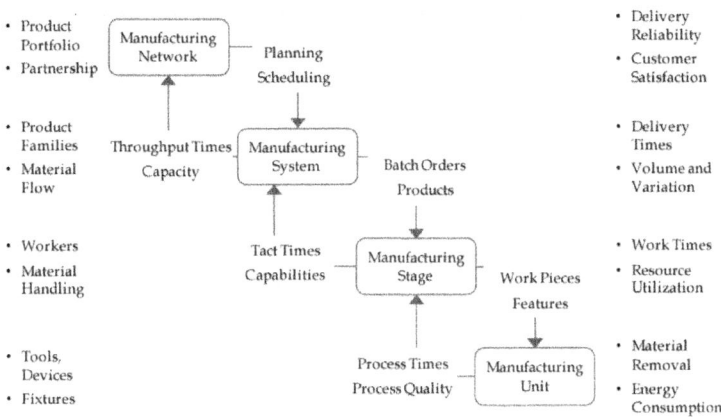

Figure. 4: Examples of structuring levels of manufacturing and their connections

Manufacturing stages are physical or logical manufacturing areas, e.g. manufacturing cells, consisting of one or more manufacturing units and their co-operation. Additionally, the manufacturing stages include internal material handling in moving the work pieces between the manufacturing units as well as buffers and stocks to hold batches of the work pieces. In manufacturing stages the focus can be on simulation for the reachability and sequences of operations as well as internal work cell layout and material handling (Kühn, 2006). Manufacturing plants are composed of manufacturing stages, warehouses for storing the products as well as internal logistics to transfer material between the stages and material storing areas. They typically correspond to factories and have customers who can be other companies or internal customers, such as an assembly plant. Typical simulation issues concern the layout design and material flow analysis as well as planning and controlling the manufacturing activities. Simulation studies on a manufacturing plant level are usually conducted using discrete event simulation (DES) including the need for and the quantity of equipment and personnel as well as evaluation of operational procedures and performance. Manufacturing networks consist of factory units, which can exist globally. One of the key differences between plants and networks is that entities in the network often belong to different companies that may have contradictory goals in their strategies. Simulation can be focused on traditional supply chain sales and delivery processes as well as to complex networked manufacturing activities, including logistical accuracy and delivery reliability of increasing product variety.

DIGITAL MANUFACTURING SUPPORT FOR MANUFACTURING ACTIVITIES

A digitally presented manufacturing system contains the information and knowledge of manufacturing entities and activities that it is reasonable to represent in a digital form. This, at its best, makes possible efficient collaboration between all the manufacturing activities and related parties. The discussion on digital manufacturing support is based on a previously developed framework for extended digital manufacturing systems (EDMS). An EDMS can briefly be defined as follows (Nylund and Andersson, 2011):

- an integrated and collaborative environment for humans, machines, and information systems to act and interact;
- to enhance the research, development and management activities of products, production systems, and business processes,
- supporting knowledge-intensive decision-making in the entirety of their lifecycles.

From Ideas to Innovative Solutions

Figure 5 represents a process from ideas and the need for change to innovative solutions. It consists of a chain of activities where the results evolve towards more precise solutions. Each phase has its enablers as inputs and the activity creates results as outputs. The results affect the enablers in the following phases of the process. The process is also iterative as it is possible to go back to previous phases in order to change or refine them. The need for change can arise, for example, from social, technological, economic, environmental, and political aspects. The changes can also derive from voluntary ideas that are seen to improve the competence of the system. If the process has not been developed previously, the current system has to be analysed to create the digital information and knowledge of what currently exists. The synthesis of the existing system and possible changes form the new requirements for the future system. The combination of feasible new possibilities and existing capabilities forms the solution principles. The results are digital entities and abstract and conceptual descriptions, including the objectives and preliminary properties of the future system. When the descriptions evolve towards a more detailed level, possible technologies can be investigated, resulting in alternative solutions. The solution alternatives can be modelled as virtual entities that include, in addition to their digital description, for example, 3D models with their own operating rules, motion, and behaviour. Combining the existing and new virtual entities forms a rough simulation model. The solution that is implemented has to be verified to make sure that the behaviour and co-operation of the entities in the system are modelled correctly. The verified simulation model can be used to run test experiments. By analysing the results from the simulation model and comparing them with known or predicted outcomes, the behaviour of the simulation model can be validated. When the simulation model is verified and validated, it can be used for manufacturing experiments. The experiments are used to analyse the behaviour of the system, and can lead towards innovative solutions.

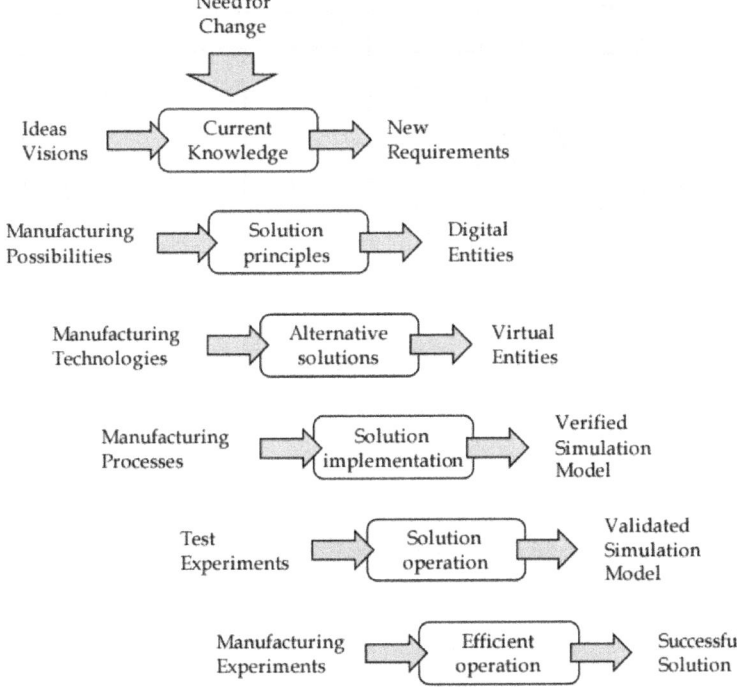

Figure. 5: The process from ideas to innovative solutions

Manufacturing Process and Flow Development

Figure 6 shows a theoretical example of process and flow development. The manufacturing process part corresponds to the process domain, presented in Figure 3, where the capabilities for a manufacturing network are developed. The part of the manufacturing flow presents the production domain in Figure 3, aiming for the right capacity and scalability of the manufacturing network to meet the customer demands. The existing capabilities are combined with new possibilities, requirements, and constraints in the production network creating the synthesis of existing and what new capabilities will be required. These derive from, for example, new possible markets, customers, and competition i.e. what is important in the future that the current capabilities cannot fulfil. The new possible capabilities are tested virtually using computer-aided technologies in connection with the digitally presented information and knowledge.

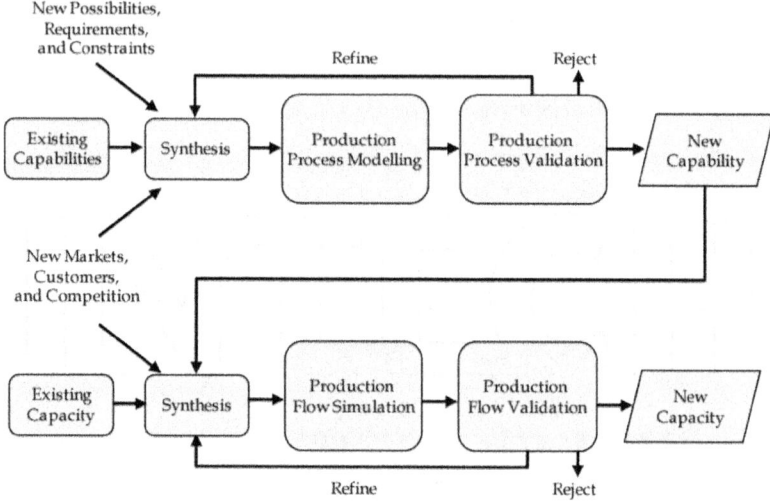

Figure. 6: Production process modelling and production flow simulation.

The resulted new capability is validated both to ensure that it does what it is supposed to do and that it meets the performance requirements, such as cost, quality, and time efficiency as well as the social and environmental issues. The production flow simulation in Figure 6 follows the same idea as the production process modelling. The new capability can add to the total capabilities of the network if something new is implemented, or change the existing capabilities if something already existing is reconfigured. It is not enough that all the needed capabilities exist. The production flow simulation aims to define how much capabilities are required to produce the changing volume and variation of customer orders at the right time. Typical areas are the controlling, planning and scheduling of the activities. To investigate the production process modelling in more detail, five categories between product requirements and resource capabilities can be recognized, see Figure 7:

- Existing capability: The capabilities exist for all of the product requirements without any need for changes to the system. The products can be manufactured as the service requests have service providers.

- Possible existing capability: At least some of the product requirements need further investigation as to whether the capabilities exist or not. The requirements are close to the existing capabilities and, using modelling and simulation, the capabilities can be verified.

- Capability after reconfiguration: There is no existing capability but it may be possible to reconfigure the system so that it has the capabilities. By modelling the reconfigured system the possibility can be verified.

- Capability after implementation: The system does not have the needed capability. It may be possible if new capabilities are added to the system. Again this can be verified using modelling and simulation.

- No capability: The result may also be that there are no capabilities and they cannot be implemented either. This leads to the need for an alternative solution, which leads to a result that fits into one of the first four categories.

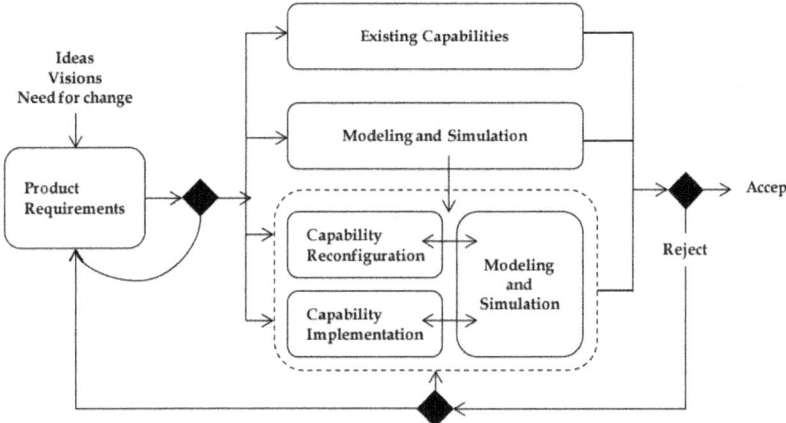

Figure. 7: Alternative outcomes of capability modelling and simulation

When it is known that the capabilities exist for all the product requirements, the efficiency of the capabilities still needs to be evaluated against factors such as cost, quality, and time. It has to be decided if the solution alternative is good enough. It can be further investigated in the capacity loop or it can be rejected and sent back to the capability loop. If all the needed capabilities exist, the capacity of the system has to be checked. The same five categories can be used in capacity evaluation. If it is known that there is enough capacity, nothing else has to be done. Modelling and simulation can be used to verify that there is enough capacity. It can also be used in capacity reconfiguration and implementation issues. Modelling and simulation of capacity has the same constraints as in the case of capabilities. The capacity for existing volume and variation still has to exist when new products are considered as an addition to existing products. In the capacity loop, the solution can be accepted or rejected, as in the capability loop. If the solution is rejected, it can be sent back to the capability loop or further back into the design requirements loop.

Manufacturing System Operation

Operation of a manufacturing system can be viewed from the time dimensions of past, present, and future. The past represents what has happened i.e. it can be said to be the digital memory of the system. The time dimension of the present, what is happening now, is used to operate the current system by monitoring the state of the system and comparing it to the desired state. The future dimension makes it possible to plan future manufacturing activities ahead and to compare different changes in strategies. Figure 8 shows the connection of the time dimensions into the operation of manufacturing systems. The past presents the data collected from the system activities when they happened. It can be used to analyse previous manufacturing activities in order to find out what happened and the reasons why it happened. In finding the root causes for phenomena, the system can learn from its past and prevent unwanted situations in the future. Rules for the autonomy of the manufacturing entities, as well as for their collaboration, can be enhanced and new rules can be created. The present here means the near future, where no major changes are planned.

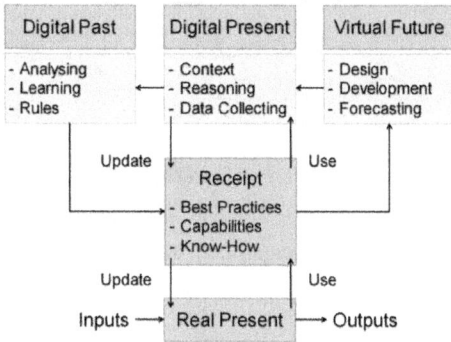

Figure. 8: Digitally co-existing past, present, and future time dimensions

It is, for example, the use of existing resources and the planning and scheduling of customer orders that have already been placed. In the present the digital and real existences co-exist. As the system operates the activities are logged, creating new history data to be analysed and to aid decision-making. The state of the real manufacturing system can be seen in the digital manufacturing system and actions can be taken with the state of the system as a starting point. The dimension of the future relies on the information and knowledge gathered from the system previously. Future design and development decisions are syntheses of existing capabilities and requirements combined with future goals and possibilities. The viewpoint of the future can be divided into tactical decisions and visions. Tactical decisions consider the near future into which the manufacturing system is heading. Future visions are similar to tactical decisions, the difference being the time horizon. The outcome of future visions is more obscure but there are more possibilities to be investigated. The information and knowledge from analysing the past, collecting data from the present, and forecasting the future is stored in the form of receipts. A receipt holds the capabilities of a system, constantly updating and refining the best practices in conjunction with human skills and know-how. The receipts are the basis of the operations in the real present, the only time dimension in the real world.

Continuous Analysis and Improvement

A manufacturing system can be seen as multiple autonomous manufacturing entities interacting and co-operating in a complex network of manufacturing activities. The activities are explained as services, which hold the information and knowledge needed to explain the manufacturing activities. It is required that the activities are known exactly, in that they are understood by all related parties. Describing the activities as services in a digital format creates a formal way to present the services. This makes possible efficient collaboration in a digital manufacturing system between entities that can be humans, machines, or information systems. The information and knowledge is kept as the autonomous property of the manufacturing entities and the communication between the entities includes only the information that is needed to fully describe the collaboration activity. The communication between the manufacturing entities is loosely based on service-oriented architecture (SOA), which consists of self-describing components that support the composition of distributed applications (Papazoglou & Georgakopoulos, 2003) enabling the autonomous manufacturing entities to negotiate and share their information and knowledge. The basic conceptual model of the SOA architecture consists of service providers, service requesters, and service brokers (Gottschalk, 2000). The roles of manufacturing entities in a digital manufacturing system based on SOA are briefly explained as follows:

- Service requesters are typically product entities when they are realized as order entities. The order entities call on the services they require to be manufactured.

- Service providers include the manufacturing resource entities which have the capabilities needed to provide the services that are requested.

- Service broker plays a role of an actor that contains the rules and logics of using the services. Its function is to find service providers for the requesters on the basis of criteria such as cost, quality, and time.

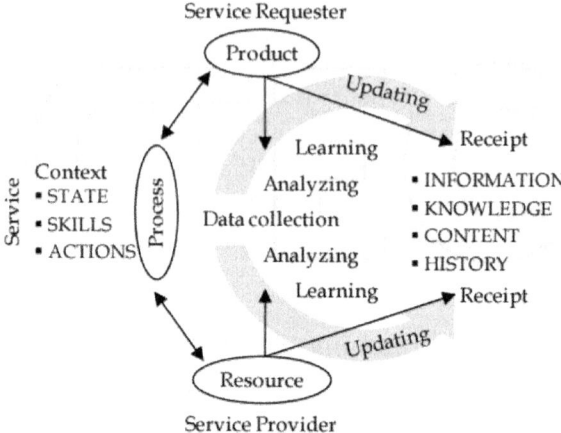

Figure. 9: An example of a service between a service provider and a service requester

Figure 9 shows an example of a service occurring in the process domain between products and resources. A service consists of two different entities, i.e. the product and resource entities having the roles of service requesters and providers. The actual service, being the manufacturing activity, is twofold, consisting of a context and receipt. The context is the environment, real or virtual, where the service takes place, whilst the receipt is the digital description of the service. The product entity requests a service, which is provided by the resource entity. The service, whether it is happening in a virtual or real environment, has a certain context that is in a certain state. The state is a basis for the actions happening during the service, and the result is based on the skills of the service provider. During the service data are collected from the process. The collected data are analyzed, forming information that is the basis for learning from the service. When something is learned, it is used to update the receipt, which will be the basis for future services. When a certain product entity uses a service provided by a certain resource entity, the data collection, analysis, learning, and updating phases include adding the same data and information to the knowledge of both entities. The knowledge of a resource entity is updated with several product entities using the services it provides. In a similar fashion, the knowledge of a product entity consists of all the services it requests. A service can be seen as a hierarchy in which a service on the upmost level divides iteratively into multiple subservices until the level on which the individual part features are requested. This means that an entity requesting a service gets information about the possible service provider entities, but it does not know how the service request is fulfilled. For example, a service request for the manufacturing of a product is a request on the macro level. The macro level service request is divided into multiple sub-services on the meso level and the meso-level service makes similar requests on the micro level. The upper level only needs the information about whether the service request can be fulfilled or not. The hierarchy of the services may be limited by the service requester as it may state special requirements for the service that limit the selection of possible providers. For example, a customer may require certain parts of ordered products to be manufactured in a specific manufacturing plant.

ACADEMIC RESEARCH ENVIRONMENT

Several of the theoretical issues discussed in this chapter have been implemented into an academic research environment of which real machinery exists in the TTE heavy laboratory, see Figure 10. The digital part of the environment has been constructed as a modular ICT architecture and the virtual part exists as simulation and calculation models. The aim of the environment is to offer a research platform that can be utilised in:

- Designing, developing and testing current and future research topics.
- Prototyping possible solutions for industrial partners in ongoing research projects.
- Utilizing it as an educational environment for university students and company personnel to introduce the latest results in the area of intelligent manufacturing.

Figure. 10: The research environment in TTE heavy laboratory

The initial version of the environment was introduced during the Tampere Manufacturing Summit seminar, which was held in Tampere, Finland, in June 2009. Since then the environment has been discussed in scientific research papers as well as in seminar and conferences.

General Description of the Research Environment

The research environment consists of typical manufacturing resources and work pieces as physical entities. The resources of the research environment, offering different manufacturing capabilities are, see Figure 11:

- Machine tools (a lathe and a machining centre) for machining operations.
- Robots for material handling and robotized machining operations.
- Laser devices for e.g. machining and surface treatment.
- A punch press, existing only virtually, for the punching of sheet metal parts. The real punch press is located at a factory of an industrial project partner company.

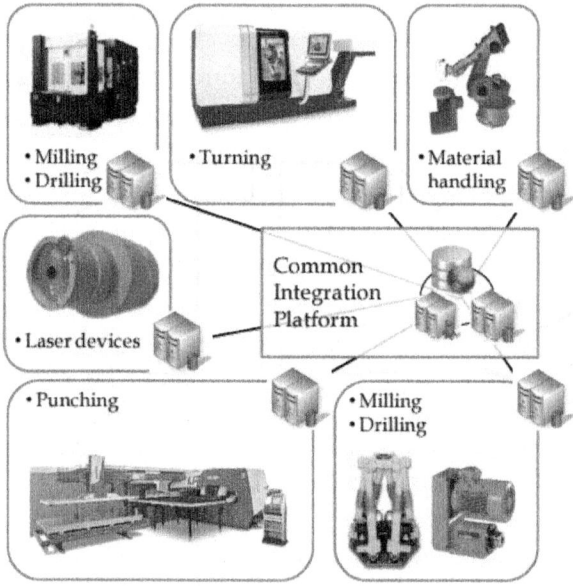

Figure. 11: The machine tools and devices of the research environment

The work pieces, which can be manufactured in the environment, are fairly simple cubical, cylindrical, and flat parts in shape. They have several parameterized features that can be varied within certain limits, e.g. dimensions (width, length, and depth), number of holes, internal corner radiuses, and sheet thickness. The main reasons for the parameterization are, firstly, that the number of different parts can be increased with the variation without having a large number of different types of parts. Secondly, the parameters can be set in a way where changing the parameters also requires capabilities of different kind i.e. different manufacturing resources are required. This gives more opportunities to compare alternative ways to manufacture the work pieces based on selected criteria, such as the cheapest or fastest way to manufacture a work piece.

Viewpoints of the Environment

The research environment can be seen from the digital, virtual, and real viewpoints. Figure 12 shows the digital, virtual, and real views of the whole research environment. The environment can be viewed from three different structuring levels; the whole environment, machining and robot cells, as well as the individual machine tools and robots. The real part of the environment exists in a heavy laboratory and is divided into two main areas, one including the robots and laser devices, and the second consisting of the machine tools. The real manufacturing entities on each structuring level have their corresponding computer models and simulation environments as their virtual parts. The information and knowledge of the environment is stored in local databases of the manufacturing entities as well as in a common Knowledge Base (KB) for the whole environment, those presenting the digital part of the environment. The actual connection is enabled by and executed via the KB, see (Lanz et al., 2008), as all communication activities use or update it. The KB is the base for the ICT-related research and development activities of the research environment. It is a system where the data of the environment can be stored and retrieved for and by different applications existing in the environment.

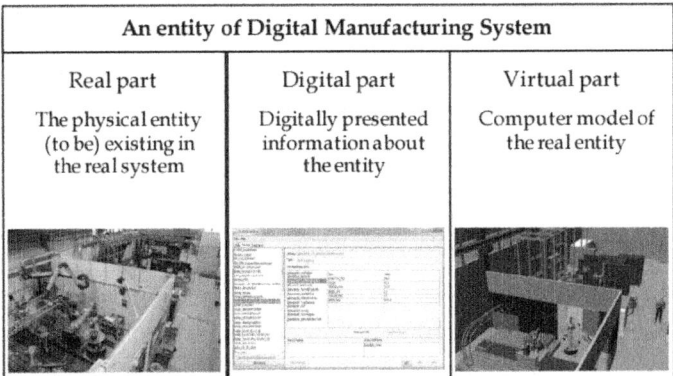

An entity of Digital Manufacturing System		
Real part	Digital part	Virtual part
The physical entity (to be) existing in the real system	Digitally presented information about the entity	Computer model of the real entity

Figure. 12: Digital, virtual, and real viewpoints of the research environment

Scenarios for Manufacturing Tests

Figure 13 presents an overall view of the process of digital, virtual and real manufacturing tests that can be performed using the research environment. The product and manufacturing information and knowledge holds what is known about the manufacturing resources of the environment and products that have been manufactured in the environment. The manufacturing methods of the resources are described as capabilities of the research environment i.e. what is known that can be manufactured within the environment. The product requirements are described similarly including all manufacturing features of products that have been previously manufactured. When the ability to manufacture a new product will be examined, firstly a CAD model of the product is required. The CAD model will be analyzed using a feature recognition property of the research environment. For each product feature, a service request is created.

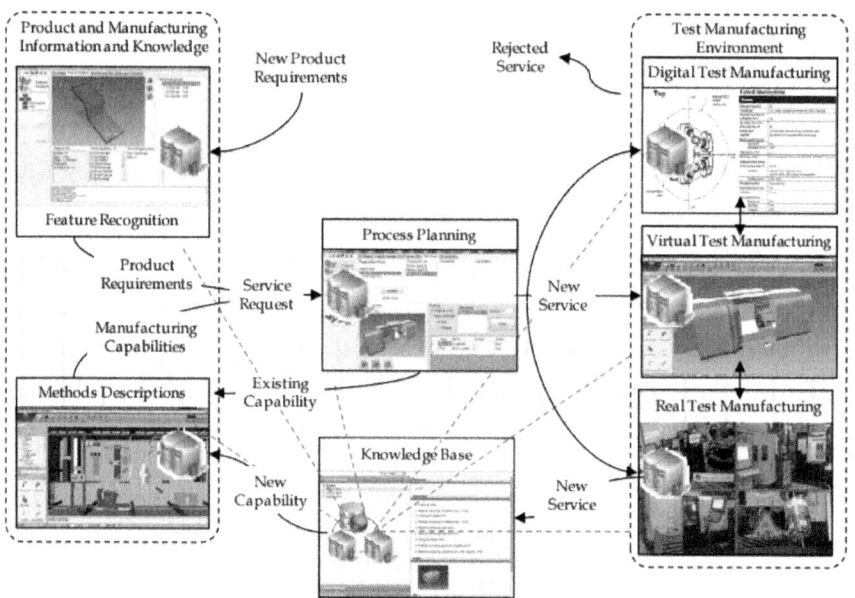

Figure. 13: An overall view of the manufacturing tests of the research environment.

The request is sent to the process planning part of the environment to compare the requirements of the new request to existing capabilities of the environment. If a suitable service exists i.e. there exists a process plan for the product feature, the result will be an existing service and no further examination is required. Otherwise, the new service request will be tested for its manufacturability. The manufacturing tests can roughly be divided into three categories being digital, virtual and real test manufacturing. The digital test manufacturing is basically comparing a set of parameterized values of the service request to the formally described capabilities of the manufacturing resources. The process is quite rapid as it is happening in a computer and no visualization or animation is required. It is the most favourable choice if time is limited. The second choice would be a virtual test manufacturing i.e. typically modelling and simulation. It requires more time as human interaction is required during the process. The time required is dependent if existing simulation models can be used or new simulation models need to be constructed. In the case, where the existing simulation models cannot be used, new ones are required to be built. The creation of a new simulation model may be to reconfigure the existing virtual system to meet the new requirements, or implementing something new into the system if the system does not have all the required capabilities or manufacturing resources. In these alternatives, the test manufacturing is still carried out with computers i.e. it does not interrupt the use of the real manufacturing resources. The real test manufacturing will be the choice if the digital or virtual manufacturing tests are not accurate enough to fully trust or understand the results gathered from the test. The real test manufacturing requires the physical resources and the time used will reduce the time for daily operations to manufacture customer orders. In some cases it is also reasonable to conduct additional tests with real manufacturing resources to reduce the risk of implementing fault processes. The responsibility of selecting, whether digital or virtual test manufacturing would be enough, is to be determined by humans, based on their skills and knowledge of the matter in hand, and has to be evaluated separately for each time a decision needs to be made. After the manufacturing tests have been conducted, the alternative is either a rejected or accepted new service. The result of rejected service could happen if the product feature cannot be manufactured within the system, or even if it could be manufactured, it is e.g. too expensive, uses too much time or does not output desired quality. In these cases, the results can be fed back to the product development to consider it the feature can be redesigned. In the case where the new service is accepted, it is added as a new capability of the environment and new process plan will be created. This will increase the known capabilities of the environment as each test manufacturing test adds new information and knowledge to the digital part of the environment, which will be available for the future test manufacturing cases.

Performance Metrics

The measurements of the manufacturing environment can be divided into direct and indirect measures. The direct measurements are achieved using the sensors and measurement devices in the environment, and the metrics can be calculated immediately. Examples of the direct measurements are:

- Process quality assurance, a real time measurement using force, acceleration, and acoustic emission (AE) sensors.

- Process stability monitoring following the electricity variation of the robot servomotor caused by the cutting forces.

- Energy consumption monitoring using a Carlo Gavazzi EM21 72D energy meter.

In the case of the indirect measurements the logged data are stored in the history section of the KR. The data can be analyzed and to create the desired performance metrics. Table 1 summarizes the performance metrics from the viewpoints of manufacturing operation, production supervising, and business management.

Table 1: Different views to utilize the performance metrics

Performance metric	Manufacturing Operator	Production Supervisor	Business Management
Cost	Continuous improvement to reduce the cost per part	Using the most cost-efficient production choices	The gain more profit and offer cheaper products to customers
Quality	To assure the manufacturing process efficiency and stability	Delivery reliability and Just-In-Time manufacturing	Improved customer satisfaction and decreased reclaims
Material consumption	To use near-net shape blank material	To reduce waste, material and energy use to meet the sustainability requirements	Meeting the requirements of legislation and expectations of the society by reducing the unwanted effects
Waste			
Energy consumption	To have real energy consumption results		
Production load and time metrics	To reduce the time per part and to update any changes in the manufacturing process times	To efficiently plan and schedule production to utilize the capacity of the system	To know how much customer orders can be placed and to give more precise delivery dates
Resource utilization			

CONCLUSION

This Chapter discussed on the possibilities of digital manufacturing to support efficient activities of designing, developing and operating manufacturing systems. A structure of individual manufacturing entities and whole systems was proposed. Describing entities of a manufacturing system as independent, yet closely related existences of digital, virtual and real enables more efficient and effective manufacturing activities from early conceptual ideas to successful solutions. Even when describing the manufacturing entities independently, they are required to be closely integrated with each other and that can be done via domains of manufacturing related activities of products, resources, and business. Again, when the entities and domains are combined, the integrated fashion should also be invested separately in different structuring levels of manufacturing, yet again closely integrated between the structuring levels By keeping the entities the same during their whole lifecycle reduces the loss of information and knowledge and enables more efficient manufacturing activities. These we discussed from several aspects i.e. a path from early ideas and needs to efficient solutions, development of manufacturing processes and flow, as well as how a system can learn from its daily operations by collecting and analysing data from the activities that can help in learning thus improving the way to do things in future. An academic research environment was discussed on how these theoretical aspects can be implemented into a manufacturing environment. As the environment is constantly developed, some of the issues have been fully implemented while some other areas remain as a future of the environment. This is due to the fact that the current and future research topics lead the development of the environment.

ACKNOWLEDGMENT

The research presented in this paper is co-financed by Tekes (the Finnish Funding Agency for Technology and Innovation), TUT Foundation and several major companies in Finland. The authors would like to thank all the colleagues involving on the development of the academic research environment.

REFERENCES

1. Ad-hoc Industrial Advisory Group. (2010). Factories of the Future PPP - Strategic Multiannual Roadmap.

2. Alavi, M. & Leidner, D.E. (2001). Knowledge Management and Knowledge Management Systems: Conceptual Foundations and Research issues. MIS Quarterly, Vol. 25, pp.107-136, ISSN: 0276-7783

3. Andersson, P.H. (2007). FMS in 2010 and beyond. Tampere Manufacturing Summit 2007, Tampere, Finland, June 2007

4. Bracht, U. & Masurat, T. (2005). The Digital Factory between vision and reality. Computers in Industry, Vol. 56, pp. 325-333, ISSN: 0166-3615

5. Chryssolouris, G., Mavrikios, D., Papakostas, N., Mourtzis, D., Michalos, G. & Georgoulias K. (2008). Digital manufacturing: history, perspectives, and outlook. Proceedings of the Institution of Mechanical Engineers Part B: Journal of Engineering Manufacture, Vol.

6. 222, No. 5, pp. 451-462, ISSN (printed): 0954-4054. ISSN (electronic):

2041-2975 ElMaraghy, H.A. (2005): Flexible and reconfigurable manufacturing systems paradigms.

7.   International Journal of Flexible Manufacturing Systems, Special Issue on Reconfigurable Manufacturing Systems, Vol. 17, pp. 261-276, ISSN (printed): 0920-6299. ISSN (electronic): 1572-9370

8.   Gottschalk, K. (2000). Web Services architecture overview. IBM developerWorks, Whitepaper

9.   Gould, P. (1997). What is agility?. Manufacturing Engineer, Vol. 76, No. 1, pp. 28-31, ISSN Gunasekaran, A. & Yusuf, Y.Y. (2002). Agile manufacturing: a taxonomy of strategic and technological imperatives. International Journal of Production Research, Vol. 40, No. 6,pp. 1357-1385, ISSN (printed): 0020-7543. ISSN (electronic): 1366-588X

10.   Jawahir, I.S., Badurdeen, F., Goldsby, T., Iyengar, D., Gupta, A., Metta, H., Stovall, C. & Ladd. C. (2009). Assessment of Product and Process Sustainability: Towards Developing Metrics for Sustainable Manufacturing. NIST Workshop on Sustainable Manufacturing, Gaithersburg, MD, USA, October 2009

11.   Jayachandran, R., Singh, R., Goodyer, J. & Popplewell, K. (2006). The design of a sustainable manufacturing system: A case study of its importance to product variety manufacturing. Intelligent production machines and systems, 2nd I*PROMS virtual

12.   international conference, ISBN 13: 978-0-08-045157-2, Pham, D.T., Eldukhri, E.E. & Soroka A.J. (eds.), pp.650-656

13.   Jovane, F. (2009). The Manufuture Road towards Competitive and Sustainable HAV Manufacturing, Tampere Manufacturing Summit, Tampere, Finland, June 2009

14.   Kidd, T. (1994). Agile Manufacturing: Forging new frontiers. Addison-Wesley Reading, 1stedition, ISBN-13: 978-0201631630, Boston, MA.

15.   Koestler, A. (1989). The Ghost in the Machine. Arkana Books, ISBN-13: 978-0140191929, London, UK

16.   Koren, Y., Heisel, U., Jovane, F., Moriwaki, T. Pritschow, G., Ulsoy, G. and Van Brussel H. (1999). Reconfigurable manufacturing systems. CIRP Annals, Vol. 48, No. 2, pp.527–540, ISSN: 0007-8506

17.   Kühn, W. (2006). Digital Factory - Integration of simulation enhancing the product and production process towards operative control and optimization. International Journal of Simulation Modelling, Vol. 7, No. 7, pp 27-39, ISSN: 1726-4529

18.   Lanz, M., Kallela, T., Velez, G. & Tuokko, R. (2008). Product, Process

and System Ontologies and Knowledge Base for Managing Knowledge between Different Clients.

19. Proceedings of the 2008 IEEE International Conference on Distributed Human-Machine Systems, pp. 608-513, ISSN: 1094-6977

20. Lapinleimu, I. (2001). Ideal Factory, Theory of Factory Planning, Produceability and Ideality. Doctoral dissertation, Tampere University of technology, Finland, Publications 328

21. Lean Enterprise Institute. (2008). Capsule Summaries of Key Lean Concepts. accessed September 15, 2011, Available from: http://www. lean.org/WhoWeAre/NewsArticleDocuments/key_lean_definitions. html

22. Maropoulos, P.G. (2003). Digital enterprise technology-defining perspectives and research priorities. International Journal of Computer Integrated Manufacturing, Vol. 16, Nos. 7-8, pp. 467-478, ISSN (printed): 0951-192X. ISSN (electronic): 1362-3052

23. Mehrabi, M.G., Ulsoy A.G. & Koren, Y. (2000). Reconfigurable manufacturing systems: Key to future manufacturing. Journal of Intelligent Manufacturing, Vol. 11, pp. 403–419,ISSN (printed): 0956-5515. ISSN (electronic): 1572-8145

24. Nagel, P. and Dove, R. (1991). 21st century manufacturing enterprise strategy. Iacocca institute, ISBN-13: 978-0962486630, Bethlehem, PA.

25. Nonaka, I. & Takeuchi, H. (1995). The Knowledge-Creating Company: How Japanese Companies Create the Dynamics of Innovation. Oxford University Press, ISBN-13:978-0195092691, Oxford, USA

26. Nylund, H., Koho, M. & Torvinen, S. (2010). Framework and toolset for developing and realizing competitive and sustainable production systems. Proceedings of the 20th International Conference on Flexible Automation and Intelligent Manufacturing, July 2010, Oakland, CA, USA, pp. 294-301.

27. Nylund, H. & Andersson, P.H. (2011). Framework for extended digital manufacturing systems. International Journal of Computer Integrated Manufacturing, Vol. 24, No. 5,pp. 446 – 456, ISSN (printed): 0951-192X. ISSN (electronic): 1362-3052

28. Papazoglou M.P. & Georgakopoulos, D. (2003). Service Oriented Computing.Communications of the ACM, Vol. 46, No. 10, pp .25-28, ISSN: 0001-0782

29. Sharifi, H. & Zhang, Z. (1999). A methodology for achieving agility in manufacturing organizations: An introduction. International Journal of

Production Economics, Vol. 62, No. 1-2, pp.7-22, ISSN: 0925-5273

30. Souza, M.C.F., Sacco, M. & Porto A.J.V. (2006). Virtual manufacturing as a way for thefactory of the future. Journal of Intelligent Manufacturing, Vol. 17, pp. 725-735, ISSN (printed): 0956-5515. ISSN (electronic): 1572-8145

31. Strauss, R.E. & Hummel, T. (1995). The new industrial engineering revisited – information technology, business process reengineering, and lean management in the selforganizing "fractal company". Proceedings of 1995 IEEE Annual International Engineering Management Conference, Singapore, June 1995, pp. 287-292.

32. Van Brussel, H., Wyns, J., Valckenaers, P., Bongaerts, L. & Peeters, P. (1998). Reference architecture for holonic manufacturing systems: PROSA. Computers in Industry, Vol.37, pp. 255-274, ISSN: 0166-3615

33. Warnecke, H.J. (1993). The Fractal Company: A Revolution in Corporate Culture, SpringerVerlag, ISBN-13: 978-3540565376, Berlin, Germany.

34. WCED (1987) Our Common Future. World Commission on Environment and Development(Brundtland Commission), Oxford University Press, ISBN-13: 978-0192820808,Oxford.

35. Wiendahl, H.-P., ElMaraghy, H.A., Nyhuis, P., Zäh, M.F. & Wiendahl H.-H. (2007). Changeable Manufacturing - Classification, Design and Operation. CIRP Annals, Vol. 56, No. 2, pp.783-809, ISSN: 0007-8506

36. Womack, J.P., Jones, D.T. & Roos, D. (1990). The machine that changed the world, Rawson, ISBN-13: 978-0060974176

Chapter 10

MIGRATING FROM MANUAL TO AUTOMATED ASSEMBLY OF A PRODUCT FAMILY: PROCEDURAL GUIDELINES AND A CASE STUDY

Michael A. Saliba and Anthony Caruana

Department of Industrial and Manufacturing Engineering, University of Malta
Malta

INTRODUCTION

A challenge that is often faced by product manufacturers is that of migration from a manual to an automated production system. The need to embark on this migration may develop from a number of different scenarios. Typical triggers for the automation of a manual production process include a need to increase competitiveness (by reducing labour costs and increasing labour productivity), a need to meet higher quality demands from the customer, an increased awareness of health and safety issues leading to the need to move human workers away from hazardous tasks, and a need to improve production efficiency parameters (e.g. reduce manufacturing lead time, improve production capacity, improve production flexibility and agility). The general strategic and/or technical aspects pertaining to the implementation of automation have been widely addressed in the literature (e.g. Asfahl, 1992; Chan & Abhary, 1996; Groover, 2001; Säfsten et al., 2007). In this work the focus is on the compilation and application of a number of standard design tools and of production system evaluation tools to facilitate and support the migration to an automated production system, in a scenario that involves a certain degree of product variety. The results are presented in the form of a set of recommended procedural guidelines for the development of a conceptual solution to the migration process, and the implementation of the guidelines in a real industrial case study. Specifically, this work addresses the situation where a manufacturer needs to investigate a potential manufacturing system migration for the assembly of a part family of products, where no or minimal product design changes are allowed. A list of procedural guidelines is proposed, in order to aid the manufacturer in analyzing the requirements for the transition, and in carrying out a conceptual design for a suitable automated manufacturing system. The guidelines are applied in the context of the industrial case study, where it is required to investigate and develop a migration plan for the assembly of three related product part families. The case study involves a relatively large manufacturing plant (about 700 employees), that produces electromechanical switch assemblies for the automobile industry. The trigger for the migration process is a significant increase in projected order volumes over a four year period, thus necessitating an increase in production capacity, as well as providing a good opportunity to obtain substantial return on investment following the implementation of advanced manufacturing technologies. The three part families under consideration are referred to as the "single gang" switches consisting of 20 variants, the "old three gang" switches consisting of 11 variants, and the "new three gang" switches consisting of 9 variants. A representative member of each of these families is shown in the illustration in Figure 1. Switch assembly is currently carried out in a mainly manual manner, with the aid of pneumatic presses to provide the required clipping forces. The projected production volumes for the current year (Year 1) and for the subsequent years are summarized in Table 1. The academic goals of this research are (i) to compile a set of guidelines for migration in a scenario of this type, and to apply the guidelines to this case study; (ii) to perform a study on assembly related similarities of the product families and to take advantage of these similarities in the automation process; (iii) to interpret the analytical results obtained from the feasibility analysis, so as to define the most suitable assembly line; and (iv) to utilize analytical tools in order to define the best possible concept at all stages of the automation.

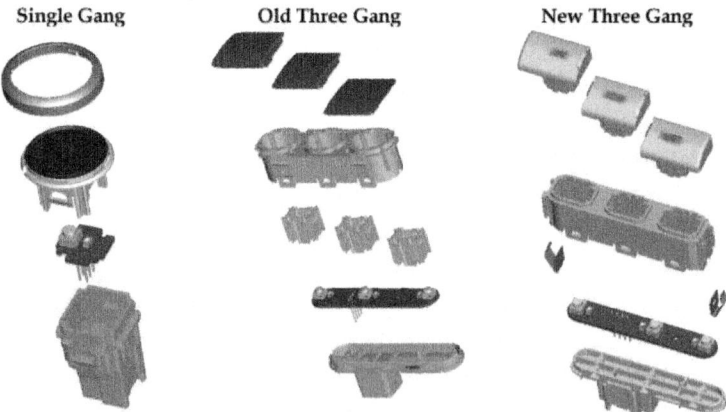

Single Gang   Old Three Gang   New Three Gang

Figure. 1: An exploded view of a representative member of each product family

From an industrial perspective, the goals are (i) to perform a cost reduction exercise on the current assembly processes; (ii) to perform a feasibility analysis of the developed concepts with respect to a number of considerations such as the assembly line balancing, cycle time reduction, production capacity and maintenance requirements; and (iii) to plan the integration of ergonomic principles in all workstations and also develop and implement safety guidelines in the process development.

Table 1: Projected annual production volumes

	Year 1	Year 2	Year 3	Year 4
Single gang switches	136,000	935,000	1,432,600	1,552,600
Old 3 gang switches	1,098,000	816,000	816,000	816,000
New 3 gang switches	0	150,000	865,000	1,495,000

LITERATURE REVIEW

In product manufacture, assembly needs to be given significant importance, since assembly related operations generally amount to 70% of the total product cost (Boothroyd et al., 2001). Therefore rationalisation of the assembly process is essential in order to optimize, mechanize and automate the activities performed, especially for assemblyintensive products. Where a new product is being developed, an Integrated Product Development (IPD) approach can be taken, whereby the marketing, design, and manufacture of the product family can be optimized and developed concurrently (Andreasen & Hein, 2000). Where the product family exists and is already in manual production, the options for design modifications may be minimal, and the development of the automated production system rests on effective categorization of the products through analytical methodologies such as group technology (e.g. Hyer & Wemmerlov, 2002), and the effective exploitation of the identified similarities. Several approaches are found in the literature to address the migration problem. Asfahl (1992) identified five phases in the implementation of automation to a currently manual process: planning, development, mock-up and test, installation, and production and follow-through. He further highlighted a number of key activities to be carried out during the planning phase: the isolation of the potential application; the identification of the project objectives; the consideration of the drawbacks; the early planning of safety aspects; the detailed documentation of the current (manual) operation; the selection of fixed versus flexible automation; and the development of a proposed layout for the system. Chan and Abhary (1996) applied an analytic hierarchy process to compare three different potential automation strategies to the existing manual plant in a case study, using a simulation approach and several evaluation criteria. Kapp (1997) introduced the "USA Principle" – Understand the existing process, Simplify the process, and Automate the process, originally intended as a guide for the implementation of enterprise resource planning (ERP), but applicable as a straightforward approach to all automation projects. Groover (2001) suggested a three-phase process whereby the individual processing stations are automated first, followed by the integration of the systems through automated handling between stations. Baines (2004) recommended a nine step approach to the manufacturing technology acquisition process: technology profiling; establishment of technology requirements; identification of a technology solution; formation of an outline business case; selection of a technology source (which may include internal development of the technology); demonstration of the technology; confirmation of the business case; implementation of the technology; and post-investment audit. Säfsten et al. (2007) suggested that manufacturing strategy development could be based on function allocation, employing a system design process that allocates various functions to either humans or machines, and optimizing the level of automation in the plant. Winroth and Säfsten (2008) further suggested an automation strategy whereby the bottom-up activities (stemming from the internal need for improvement) and the top-down activities (stemming from the market requirements) are both taken into consideration in the optimization exercise. Most of the technical literature on the development of automation systems focuses on the process that needs to be automated and/or on the product that needs to be manufactured, but less so on the production system as a product in its own right. Thus, a number of design methods that have become a mainstay in product design are not normally prescribed in a systematic way to the development of integrated automation systems. Design methods include tools and techniques such as product design specification (PDS), morphological charts, decision matrices, and failure modes and effects analyses (FMEA) (e.g. Dieter & Schmidt, 2009). This is the research gap that has been identified in this work, and the results reported in this chapter attempt to bridge this gap by drawing on these design methodologies, and prescribing them side by side with other conventional developmental steps, in order to optimize the conceptual design process for an automation system, in an environment of variety in the products that need to be manufactured. The contribution of this work is further extended to include a detailed illustration of the step by step application of the procedural guidelines to a complex industrial case study.

THE PROCEDURAL GUIDELINES

A systematic approach to the conceptual design of a new, automated manufacturing system when migrating from manual assembly of a product part family, where little or no change in product design is allowed, is presented in Table 2. The proposed list of procedural guidelines and development tools given in the table has been compiled on the general basis of the discussion given in section 2, and the developmental steps are intended to be applied sequentially, for the case of a high production volume environment in the presence of product variety. The guidelines are intended to cover the early technical and feasibility studies. Thus it is pre-assumed that the company has already taken a strategic decision to analyze the selected manual process with a view to implementing automation (if feasible), and the guidelines lead to the end of the conceptual design phase but do not address any part of the embodiment design phase or of the development of test or prototype hardware. In the general literature, it has been estimated that about 75% of the product cost is normally already committed by the end of the conceptual design stage (Ullman, 1997), and in this work the research boundary has been set to address this critical phase of product development. It is emphasized once again that it is the production equipment that is being referred to and considered as the "product" in the context of the previous sentence, rather than the objects (products, or product part family) that will be manufactured by the equipment. In the following section, the use and implementation of the procedural guidelines is illustrated in the context of the industrial case study, for the development of the conceptual design of a new manufacturing system for the assembly of the three families of automotive switches. The results for each step are summarized, presented and discussed.

IMPLEMENTATION OF THE PROCEDURAL GUIDELINES: A CASE STUDY

Analysis of the Product Family Designs

In a manufacturing environment, parts having similar geometric shapes and sizes or similar processing steps may be grouped into part families, in order to facilitate their design and/or their production. This manufacturing philosophy is referred to as group technology. Thus, parts within a particular part family will all be uniquely different, however they will have enough similarities to classify them together as one group (e.g. Groover, 2001). In the present case study, the product designs are fixed, and therefore the application of group technology principles is intended to facilitate the production of the parts. Parts classification systems are normally based either on similarities in design attributes, or on similarities in manufacturing attributes, or on similarities in both design and manufacturing attributes; however other types of similarity may also be used. Hyer and Wemmerlov (2002) identify nine criteria that may be used to classify parts, based on similarities in product type, market, customers, degree of customer contact, volume range, order stream, competitive basis, process type, and/or product characteristics.

Table 2: The procedural guidelines

1.	Analyze the current product family design, with a view to understanding clearly all similarities and variations between the members of the family and between their components.
2.	Analyze the current assembly processes and the existing assembly line(s), with a view to understanding the processes, and identifying drawbacks and opportunities for simplification.
3.	Perform a capacity analysis based on the current set-up, with a view to understanding and defining current capabilities and limitations.
4.	Draw up a product design specification (PDS) chart for the new production system, with a view to defining the requirements and wishes for the new system.
5.	Perform a group technology (GT) analysis, with a view to confirming/revising the parts classification in the context of automated manufacture.
6.	Create precedence diagrams for the process, with a view to understanding the various ways in which assembly operations can be carried out.
7.	Set up a morphological chart for the overall operation, with a view to identifying various alternatives for carrying out the various process steps.
8.	Draw up a number of different layouts at the conceptual level, with a view to identifying different alternatives for the assembly.
9.	Perform a provisional analytical study of each of the concepts, based on various criteria such as achievable cycle times, quality, shop floor area, and flexibility.
10.	Draw up a decision matrix to select the most suitable concept.
11.	Carry out a process failure modes and effects analysis (PFMEA), with a view to identifying and addressing failure mechanisms.
12.	Perform a safety analysis, with a view to identifying and addressing production hazards.
13.	Perform an ergonomic analysis, with a view to optimizing the production system with respect to interactions with human workers.
14.	Carry out a new capacity analysis for the new system, with a view to quantifying the achievable capabilities through automation.
15.	Perform a provisional return on investment analysis, with a view to quantifying provisionally the projected savings and break even times upon implementation of the new system.

In this case study, a preliminary analysis based on product design strongly indicated that the parts fell into three natural groupings as shown in Figure 1 above, based on the overall features of their geometries. All of the 20 variants of single gang switches included a socket, a printed circuit board (PCB), and a push button, as shown in Figure 2(a). The PCB included one or more coloured light emitting diodes (LEDs) as required. Some of these variants included one or more of three additional parts: a chrome ring to provide a different aesthetic finish (such as in the switch shown in Figure 2(b)), a light shield for variants that had two different graphics on their front face (to prevent light leakage between graphics), and a jewel

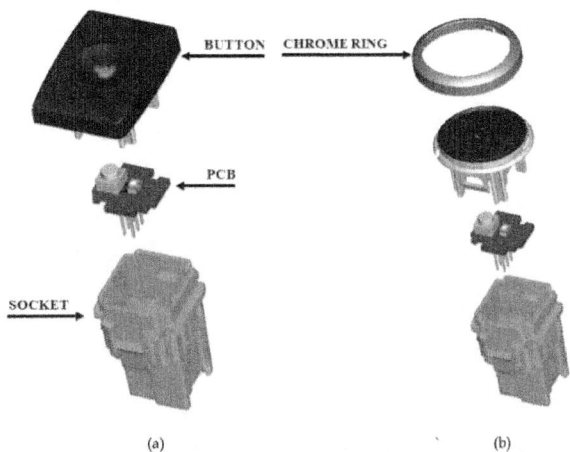

(a)   (b)

Figure. 2: Single gang switches. (a) a basic variant, (b) variant with a chrome ring

(press fit into the button) to transmit light from the LED to the surface of the button. All of the 11 variants of old three gang switches included a socket, a PCB, one or more sliders, a three unit housing, and three buttons as shown in Figure 3(a). The differences between the variants were defined by the types and combinations of buttons (functional, display, or blank). Functional buttons require a slider, in order to actuate a tact switch on the PCB, and also have a graphic display. Display buttons have only a graphic display (illuminated by an LED on the PCB), and blank buttons have no function or display. The nine variants of new three gang switches have a sleeker design, and use metal clips to attach to the dashboard of the vehicle (see Figure 3(b)).

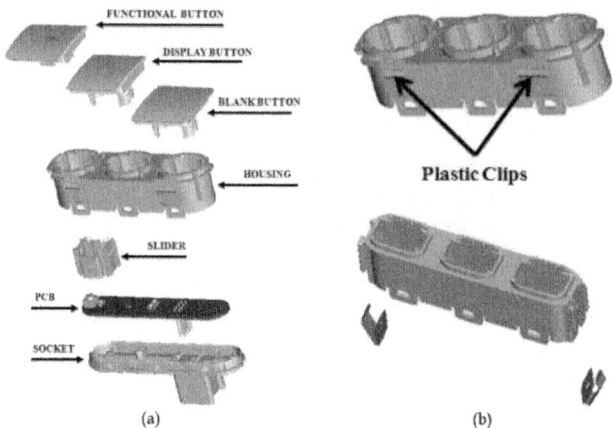

(a)   (b)

Figure. 3: (a) A variant of old three gang switch, (b) clipping mechanisms (top – old three gang; bottom – new three gang)

Analysis of the Current Manufacturing Processes

The layouts of the existing production lines are illustrated in Figure 4, with the old three gang switches and some of the single gang switches manufactured in Cell 1, and the new three gang switches and the rest of the single gang switches manufactured in Cell 2. Due to limitations in the end of line testing steps, only one model of switch can be assembled on each cell at any one time, and substantial set-up times are associated with the change over between batches of different models of switch.

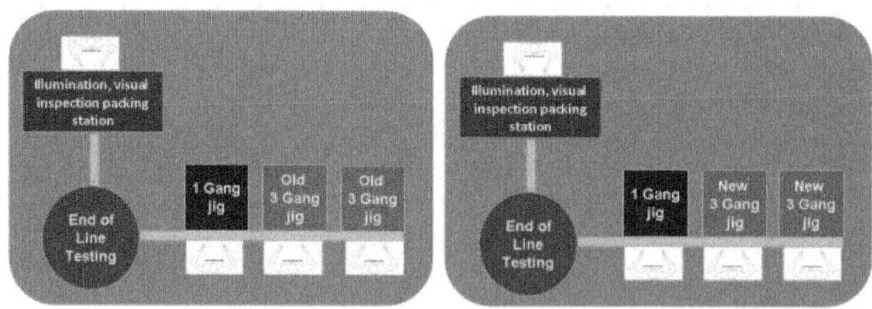

Figure. 4: (a) Layout of Cell 1, (b) layout of Cell 2

The labour intensive nature of the assembly process for the single gang switches is illustrated in Figure 5 and Figure 6. The operator reaches for one socket from the silo, and one PCB from the tray and places the socket over the

PCB in cavity (1). If a chrome ring or light shield is required for the switch being assembled, the operator reaches for the chrome

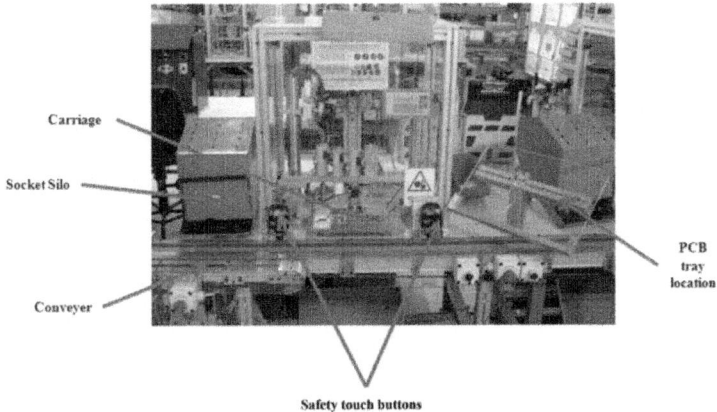

Figure. 5: Assembly jig for single gang switches

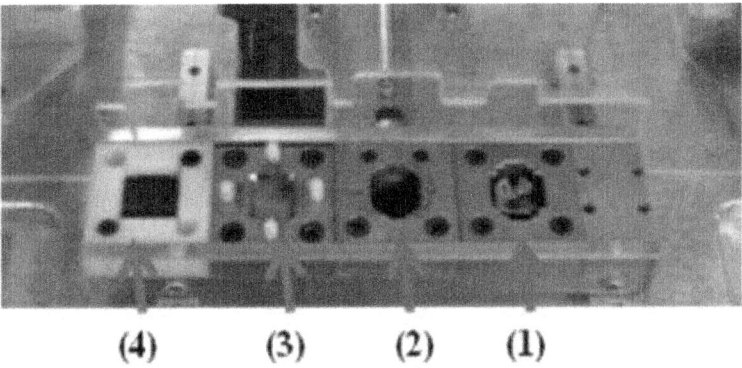

Figure. 6: The four cavities on the single gang switch assembly carriage

ring or light shield and for the button and performs manual alignment. A chrome ring and button sub-assembly is placed in cavity (3), while a light shield and button sub-assembly is placed in cavity (4). The operator then presses the two safety touch buttons placed on the sides of the assembly jig simultaneously. The carriage moves inside the jig and two pneumatic cylinders clip the socket and button sub-assemblies. Subsequently, the carriage moves outside the jig and the operator picks up the socket sub-assembly from cavity (1), rotates it and places it inside cavity (2), whilst placing the button sub-assembly on top of it. The operator also loads the parts for another socket and button sub-assemblies so that during the jig operation, three clipping processes are performed simultaneously. The operator then presses the two safety buttons simultaneously and the carriage moves inside the jig and a pneumatic cylinder clips the two sub-assemblies together. When the carriage moves outside the jig, the operator removes the switch and places it on the conveyor. For all the switches that require no chrome ring or light shield, the operator reaches for the button only when the socket sub-assembly has been placed in cavity (2). Assembly of the three gang switches is somewhat more complex due to the greater number of parts, however the nature of the operations is similar. The end of line testing is performed via a fully-automated four-station indexing table. The four stations are (i) a loading station which loads the assembled switch from the conveyor on to the station using a pneumatic pick and place device; (ii) a testing station where the switch is subjected to electrical and force testing, and (for the three gang switches) a barcode label is read; (iii) a camera and laser station where LED illumination intensity and graphic orientation is inspected, and where the customer part number and date code are engraved by laser; and (iv) an unloading station that transfers the switch, using a pick and place device, to a separate conveyor for final inspection, or onto a reject bin. The final inspection and packaging workstation is fully manual. Here the operator ensures that no scratches, dents or other defects are present on the button's surfaces; checks the integrity of the clipping features; verifies that the terminals are not bent; and ensures that the correct laser marking and date code have been used. Conforming switches are subsequently packed in the respective packaging, whilst non-conforming switches are disposed of in the reject bin.

Capacity Analysis for the Current Set-Up

A capacity analysis was performed in order to quantify the number of switches that can be assembled and tested using the existing production lines. This was done by measuring the time required for every assembly process step of each switch variant, and by analyzing the cycle times and projected production volumes for each individual variant for the four years under consideration. In this respect, the cycle time is defined as the time interval for the completion of one complete production unit. In the case study considered, the cycle time was taken to be the longest time from among the three operations performed, i.e. assembly cycle time (the time taken to assemble one full switch), testing cycle time (the time taken to complete the longest testing step adding the indexing time of the table), and finishing cycle time (the time taken to inspect and pack). Equipment availability was assumed to run at 85%, which is due to (i) one product changeover of 15 minutes per shift, resulting in a 3.33% loss; (ii) an allowance of 30 minutes per shift for maintenance activities, including 15 minutes for breakdowns and 15 minutes for planned preventive maintenance, resulting in a 6.67% loss; (iii) personnel related stoppages of 10 minutes per shift resulting in a 2.22% loss; and (iv) process yield running at 97.5%. The number of shifts required to cater for these volumes could thus be calculated using an 85% equipment availability, with 7.5 operating hours per shift, for five days a week and 48 weeks per year. The results of the capacity analysis are summarized in Table 3. Due to current layout constraints only two product families can be tested in parallel, and this means that the permissible total number of daily shifts is six. As can be seen in the table, during fiscal years 3 and 4, the total output cannot be reached because the number of daily shifts required is not achievable. In addition to this, during fiscal year 4 the number of daily shifts required to achieve the required new three gang switch volumes is 3.11 which is not achievable, since new three gang switches can only be assembled and tested on cell 2. These results pointed out the need of improving the current layouts so as to cater for the required volumes.

Table 3: Number of shifts required to cater for the projected volumes using the existing production lines

| Fiscal Year | Number of Daily Shifts Required | | | Total Number of Daily Shifts Required |
	Single Gang	Old Three Gang	New Three Gang	
Year 1	0.30	2.55	0.00	2.85
Year 2	2.28	1.97	0.32	4.57
Year 3	3.38	1.97	1.81	7.16
Year 4	3.75	1.97	3.11	8.83

Product Design Specification Chart

A PDS chart contains a detailed list of requirements that the final product must fulfil, and is drawn up prior to starting the actual design. The aim of the PDS is to encompass all of the required information for a successful solution design and to ensure that the needs of the user are achieved. The PDS also lists a number of wishes, which are specifications that are not essential for the success of the project. These wishes however give the project a competitive edge and increase the potential benefits gained through its implementation. A PDS chart was created for this project, listing all of the specifications that should be taken into account, when designing the required improvements on the switch manufacturing cells. The section of the chart dealing with the performance criterion of the production equipment is shown in Table 4. The other criteria that were considered were target product cost, required service life, serviceability, safety, environment, size, ergonomics, materials, transportation, manufacturing facilities, appearance, quality and reliability, personnel requirements, product lifespan, documentation, and commissioning. Examples of design wishes (not shown in Table 4) include the minimization of shop floor space occupied by the equipment, the use of inexpensive (but reliable) materials, and the ability to manufacture the equipment in house.

Table 4: A section of the PDS chart

Specification	Requirement	Need
Performance criterion		
Ability to assemble, test and finish the three switch families.	✓	
Ability to cater for the projected volumes, with an excess capacity of 15%.	✓	
An assembly cycle time for the new three gang switches of less than 10 seconds.	✓	
An assembly cycle time for the single gang switches of less than 8 seconds.	✓	
Ability to test single gang and three gang switches simultaneously.	✓	

Group Technology Analysis

In this case study, the preferred criterion for parts classification was found to be that based on product characteristics, since the parts fell into three natural groupings as discussed in section 4.1. The characteristics and variations of each of the three part families were analyzed in detail, with a view to confirming this classification and to prepare for the detailed technical design phase of the project. An analysis of the constituent parts of the single gang switches produced the following results: Button – there are different types of button, due to different customer requirements, mainly in terms of graphic design, shape and the type of surface. However all the buttons in the switch family have common guiding and clipping features. Socket – two types of sockets exist with the main geometric difference being the position of the foolproof feature as shown in Figure 7. A second difference is in the socket colour, where type A is black and type B is grey. PCB – there are different types of PCB having different profiles and different location of the electrical components. Chrome Ring – there is only one type of chrome ring. Jewel – the jewel needs to be aligned with the surface so it must have the same shape as the surface of the button. Two types of jewel exist that correspond to two types of surface. Light Shield – there is only one type of light shield. Similar analyses were carried out for the old three gang and new three gang switches. The part variations associated with all three product families are summarized in Table 5.

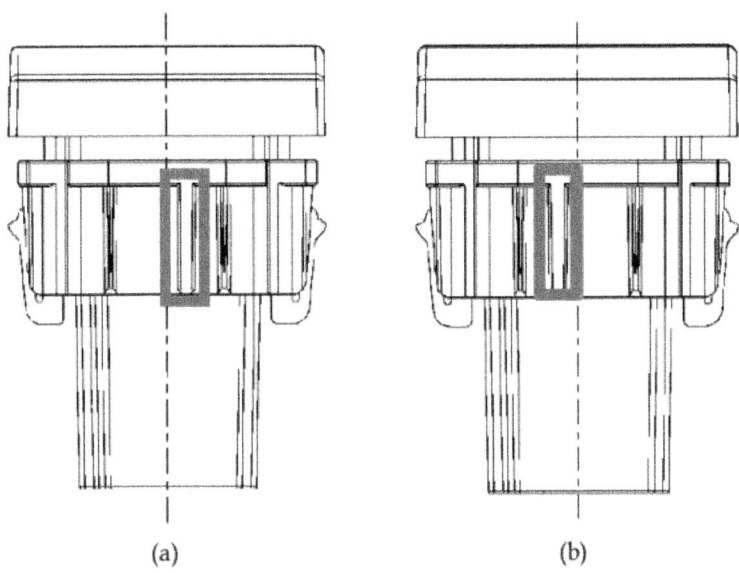

(a)   (b)

Figure. 7: The two types of socket for the single gang switch. (a) Type A, (b) Type B

Table 5: Part variations for the three product families

Single gang switches		Old three gang switches		New three gang switches	
Constituent part	*No. of Variants*	*Constituent part*	*No. of Variants*	*Constituent part*	*No. of Variants*
Button	15	Button set	10	Button set	8
Socket	2	Socket	5	Socket	2
PCB	12	PCB	9	PCB	5
Chrome ring	1	Housing	1	Housing	1
Jewel	2	Slider	1	Metal clip set	1
Light shield	1				

From the results, it can be seen that there are substantial differences between the three sets of products, in terms of the gross geometries and of the constituent parts. In particular, it is noted that a different cavity is required for each of the three families (the geometric differences between the different types of socket within each switch family are minor, and in each case can be catered for by the same cavity). At the same time, the components that constitute each product family allow for ease of automation, since there are only a small number of variations for the parts. The analysis therefore confirmed the classification of the switches into three distinct product families as indicated in Table 5.

Precedence Diagrams

The generalized manufacturing process flow chart for each switch, as extracted from the description given in section 4.2, is illustrated in Figure 8(a), and consists of three major steps. Step 1 involves the assembly of the switch. Step 2 involves the testing of the switch (force, electrical, and illumination testing) and laser marking. Step 3 involves a visual inspection of the switch and final packaging.

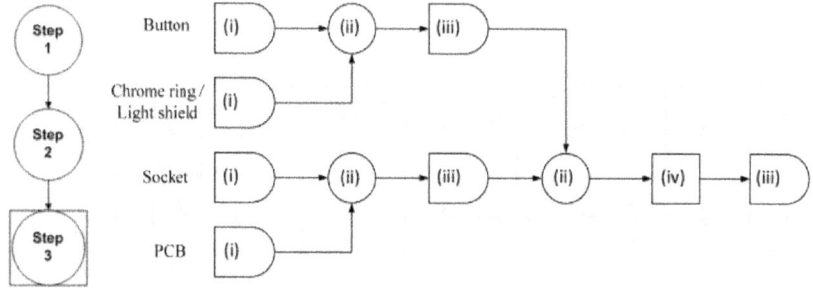

Figure. 8: (a) General process flow chart for each switch, (b) Precedence diagram for the assembly of a single gang switch

The precedence diagram for the assembly process (i.e. for Step 1) for the single gang switch is shown in Figure 8(b). In the figure, the shapes labelled (i) represent temporary storage, the circles labelled (ii) represent the complex operation "bring sub-components together, align, and clip", the shapes labelled (iii) represent delays, and the square labelled (iv) represents a brief visual verification that the clipping has been carried out correctly. For this particular case study it was not possible to simplify these steps any further. The precedence diagrams for the old three gang and the new three gang switches were extracted in a similar manner. The three diagrams served to guide the development of the morphological chart, and the generation of alternative conceptual solutions, described in the next two sections.

Morphological Chart

A morphological chart is an analytical tool which aims at finding all theoretically conceivable solutions to a problem (Roozenburg & Eekels, 1995). It provides a visual way of capturing the required product functions and of exploring possible different solutions that may exist for each product function. The chart facilitates the presentation of these solutions and provides a framework for considering alternative combinations of the individual function solutions. The main functions associated with the problem at hand were identified to be the (i) transfer system between stations, (ii) part orienting mechanism, (iii) part feeding mechanism, (iv) handling mechanism, (v) gripping mechanism, (vi) part inspection system, and (vii) part packaging. The morphological chart is shown in Table 6.

Table 6: Morphological chart for the new manufacturing system The selected solutions are highlighted

Function	Option 1	Option 2	Option 3	Option 4
Transfer System	In-line indexing system	In-Line indexing system with return carriers in the vertical plane	Rotary indexing system	Pallet system
Part Orienting	Vibratory bowl feeder	Magnetic rotary feeder	Machine vision system coupled with a robotic arm	Manual
Part Feeding	Vibrating conveyor	Linear feeder	Horizontal belt conveyor with passive guides	Manual
Handling System	Pneumatic pick and place	Electric pick and place	Robotic arm	Manual
Part Gripping	Vacuum suction	Magnetic gripping	Pneumatic grippers: radial, 3-point and angular	Manual
Part Inspection Systems	Machine vision system	Colour sensor	Human visual inspection	
Packaging System	Robot based system	Customized automation	Manual	

Solution selection was made on the following bases: Transfer system – The projected high production volumes necessitate a low indexing time, and if a pallet system is used this would only be achievable by using a very large number of pallets. Thus the manufacturing cost of the system would increase due to the large number of cavities required. A rotary indexing table reduces maintenance interventions, since maintenance requirements are less compared to that of an in-line indexing system. Part orienting – A vibratory bowl feeder can provide the required output, and is the cheapest and most reliable solution among the four options considered. Part feeding – High part feeding accuracy is required in order to ensure correct operation of the system and this accuracy can be achieved through the use of linear feeders coupled with vibratory bowl feeders. The use of vibration conveyors or of passive guides cannot achieve the required accuracy. The manual option is expensive. Handling system – The transportation of the part between two fixed positions can easily be achieved by a pneumatic handling system, which is the cheapest alternative among the four considered. Part gripping – A system based on vacuum suction could be used, however for its full implementation a number of intricate vacuum heads would need to be designed, thus substantially increasing the cost of the system. A magnetic system cannot be used for all components, since most of the components are made of plastic. A manual system is an expensive option and therefore pneumatic grippers with specifically designed jaws were selected. Part inspection system – Visual inspection is required at the end of line testing stage (which is already automated, and uses a machine vision system) and at the final inspection station. Due to the complex nature of the final inspection it was determined that this could only be carried out reliably by human operators. Packaging system – Since the final inspection is manual, the preferred option would be to have the human operator package the completed switch after inspection.

Concept Generation

Overview of Proposed Concepts

Four different concept layouts were generated to address the problem. The first concept involves automation of the single gang switch assembly, and relocation of all assembly of this switch to Cell 1. Assembly of the new three gang switch would also be automated, and Cell 2 would be dedicated to this process. The second concept involves the retention of the present, labour intensive, assembly processes, but with the incorporation of an additional station to Cell 2 for the assembly of new three gang switches, to meet the projected production volumes. The third concept is a compromise between the approaches of the first and second concepts, and involves the retention of the present, labour intensive, assembly processes for the single gang switch and for the old three gang switch, and the transfer of the single gang switch assembly station from Cell 2 to Cell 1. Cell 2 would be dedicated to the automated production of the new three gang switches as in Concept 1. The fourth concept involves the combination of all production processes into a single cell, and automating the assembly of the single gang and of the new three gang switches. It is noted that due to the fact that the production volume of the old three gang switches is expected to decrease, automation of the assembly process for these switches is not recommended under any of the proposed concepts. The four concepts are presented in greater detail in the following sections.

Concept 1

The proposed layout for Cell 1 under this approach is shown in Figure 9. The cell consists of an indexing table used for the assembly of the single gang switches, two manual jigs used for the assembly of the old three gang switches and a testing indexing table which can test the two different switches simultaneously. Linear conveyors transfer the switches from the assembly stations to the testing station. An operator loads the PCB and button on Station 1 of the loading indexing table. The work carrier of this indexing table will be sub-divided into two sections, one holding the PCB and one the button. These two parts are bought-in parts presented to assembly in painting jigs or trays and therefore automation of the loading function would require a tray changing mechanism and an x-y-z pick and place device. The

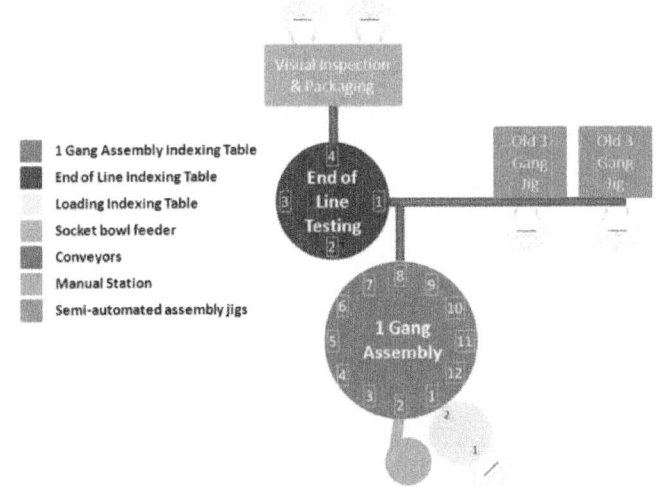

Figure. 9: Proposed Layout for Cell 1 (Concept 1)

initial cost required to create these subsystems would be much higher than the operational cost of one operator who would still be required to attend the machine, and their implementation is therefore not recommended. The parts are then automatically transferred onto a twelve station indexing table. The work carrier on this indexing table will be subdivided into three sections, namely cavities 1, 2 and 3. The proposed stations for the indexing table are listed in Table 7. The twelve station indexing table was chosen because one is already available at the company, thus reducing the initial cost required. This results in five free stations which can be utilized for future improvements of the layout. The fully assembled single gang and old three gang switches are unloaded onto conveyors which transfer them to the end of line indexing table, based on the current automated system. Table 8 lists the four stations of the fully automated indexing table. The loading station will either pick up one switch type or both switch types, depending upon the switch being available at the conveyor, since old three gang and single gang assemblies would not be synchronized. A new anti-mixing part inspection system, based on machine vision, would need to be incorporated into the end of line testing station, to distinguish between the two switch families.

Table 7: Single gang switch assembly stations (Concept 1, Cell 1)

Station	Description
1	Loading of button onto cavity 3 and PCB onto cavity 1
2	Loading of socket onto PCB
3	Clipping of socket to PCB (sub-assembly 1)
4	Turning of socket and placing onto cavity 2
5	Free station
6	Loading of button onto sub-assembly 1
7	Clipping of button with sub-assembly 1
8	Unloading
9	Free station
10	Free station
11	Free station
12	Free station

Table 8: End of line testing stations (all concepts, all cells)

Station	Description
1	Loading
2	Force and electrical testing
3	Camera test and laser mark
4	Unloading

The proposed layout for Cell 2 consists of a semi-automated twelve station indexing table, used for the assembly of new three gang switches, as shown in Figure 10. The work carrier on this indexing table will be sub-divided into three sections, namely cavities 1, 2 and 3. The socket, clips and housing are oriented via vibratory bowl feeders and automatically loaded on the respective stations. The PCB is manually loaded on a conveyor which is then automatically transferred onto the socket in station 2. Table 9 lists the proposed operations to be performed by each station. The PCB and buttons are manually loaded as in Cell 1. In station 3, the sub-assembled components are unloaded onto a conveyor and the same operator loading the PCB, adds a label to the sub-assembly. Subsequently a second operator adds the three buttons in their corresponding position and places the switch onto a third conveyor. The clipping of the buttons is performed via an automatic clipping and preactuation station and finally the switch is loaded onto the testing indexing table. The end of line indexing table is similar to the one on Cell 1.

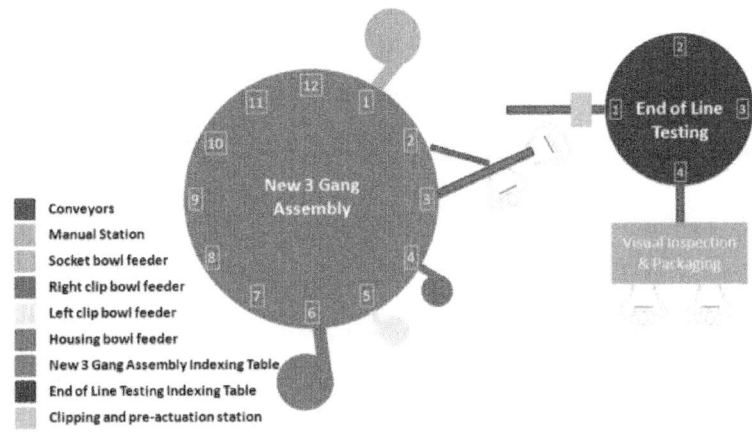

Figure. 10: Proposed layout for Cell 2 (Concept 1)

Table 9: New three gang switch assembly stations (Concept 1, Cell 2)

Station	Description
1	Loading of Socket in cavity 1
2	Loading of PCB onto socket in cavity 1 - (sub-assembly 1)
3	Unloading of sub-assembly 3 onto conveyor from cavity 2
4	Loading of right clip in cavity 3
5	Loading of left clip in cavity 3
6	Loading of housing onto clips in cavity 3
7	Clipping of housing with clips – (sub-assembly 2)
8	Loading of sub-assembly 2 onto sub-assembly 1 in cavity 1
9	Clipping of sub-assembly 2 onto sub-assembly 1 - (sub-assembly 3)
10	Transfer of sub-assembly 3 onto cavity 2
11	Free Station
12	Free Station

Concept 2

The proposed layouts under this approach are shown in Figure 11. This concept entails the incorporation of an additional manual assembly jig to Cell 2 that would be used for the assembly of new three gang switches. This is achieved by modifying the conveyor currently used to transfer the assembled parts from the jigs to the testing indexing table, so as to cater for the additional jig. Changes are also proposed to the testing program of both end of line testers, so as to reduce the testing time required. A new operator is required for the visual inspection and packaging station of Cell 2, so as to cater for all the switches being assembled. Two operators would thus be dedicated to new three gang switches, and one to single gang switches. This concept is a labour intensive concept which however requires less initial investment due to the fact that only minor modifications are required to the existing structure. The projected production volumes can however still be met through this layout.

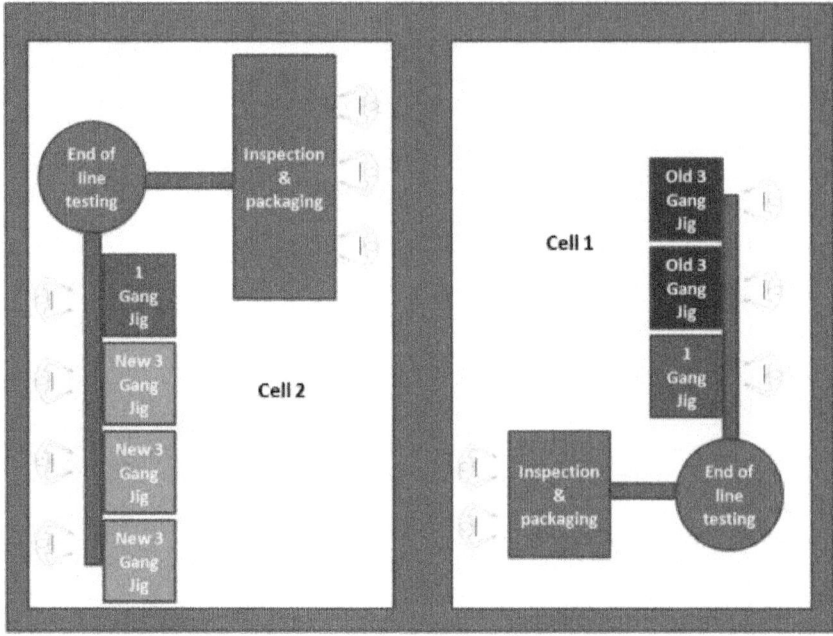

Figure. 11: Proposed layouts for Cell 1 and Cell 2 (Concept 2)

Concept 3

The layouts of the cells under this approach, as described in section 4.8.1, are shown in Figure 12. Modifications are required to the testing program of both cells so as to reduce the testing cycle time.

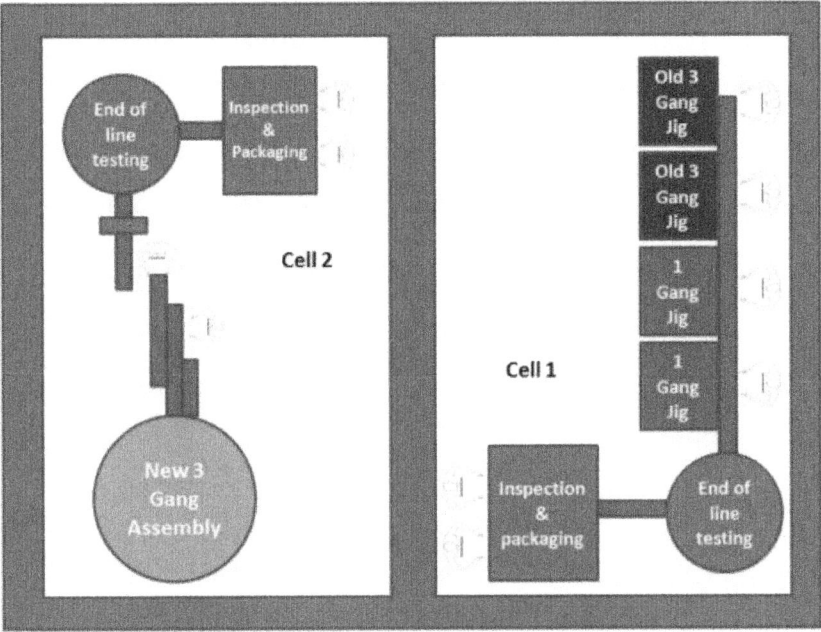

Figure. 12: Proposed layouts for Cell 1 and Cell 2 (Concept 3)

Concept 4

The fourth concept consists of one indexing table that is used for the assembly of both the single gang and the new three gang switches, as shown in Figure 13. A 20-station indexing table is used having work carriers divided into six sections, where three sections (cavities 1, 2 and 3) are used for new three gang switches and the other three (cavities 4, 5 and 6) are used for single gang switches. The main difference between this layout and the one proposed in Concept 1 involves the combination of the two assembly lines. The orienting, feeding, loading and clipping mechanisms are the same as those proposed in Concept 1. The proposed stations are listed in Table 10, where it can be seen that the first twelve stations are dedicated to the new three gang switches and the remaining eight stations to the single gang switches. There would be a total of three free stations. The end of line testing systems are the same as those proposed in Concept 1.

Table 10: Assembly stations for the single gang and new three gang switches (Concept 4)

Switch	Station	Description
New Three Gang Switch Assembly	1	Loading of socket in cavity 1
	2	Loading of PCB onto socket in cavity 1 - (sub-assembly 1)
	3	Unloading of sub-assembly 1 onto conveyor
	4	Loading of right clip in cavity 3
	5	Loading of left clip in cavity 3
	6	Loading of housing onto clips in cavity 3
	7	Clipping of housing with clips – (sub-assembly 2)
	8	Loading of sub-assembly 2 onto sub-assembly 1 in cavity 1
	9	Clipping of sub-assembly 2 onto sub-assembly 1 - (sub-assembly 3)
	10	Transfer of sub-assembly 3 onto cavity 2
	11	Free station
	12	Free station
Single Gang Switch Assembly	13	Loading of button onto cavity 6 and PCB onto cavity 4
	14	Loading of socket onto PCB in cavity 4
	15	Clipping of socket to PCB (sub-assembly 4)
	16	Turning of sub-assembly 4 and placing onto cavity 5
	17	Free station
	18	Loading of button onto sub-assembly 4
	19	Clipping of button with sub-assembly 4
	20	Unloading

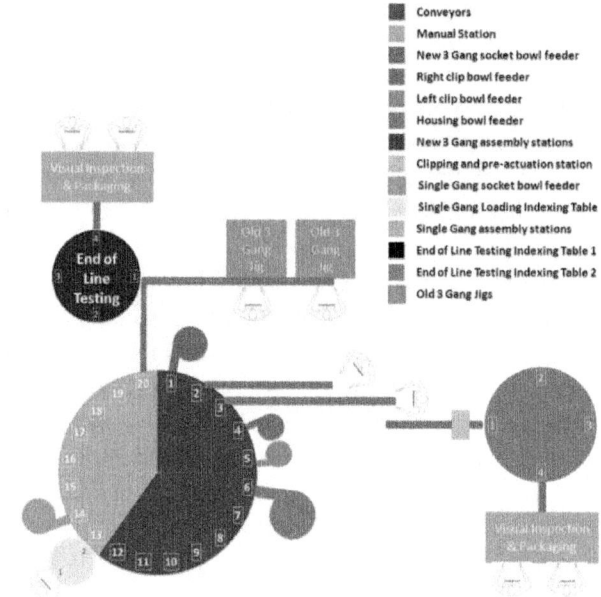

Legend:
- Conveyors
- Manual Station
- New 3 Gang socket bowl feeder
- Right clip bowl feeder
- Left clip bowl feeder
- Housing bowl feeder
- New 3 Gang assembly stations
- Clipping and pre-actuation station
- Single Gang socket bowl feeder
- Single Gang Loading Indexing Table
- Single Gang assembly stations
- End of Line Testing Indexing Table 1
- End of Line Testing Indexing Table 2
- Old 3 Gang Jigs

Figure. 13: Proposed layout for combined assembly in a single cell (Concept 4)

Provisional Analytical Studies

Overview of Analysis

The proposed concepts were analyzed with respect to a number of parameters, namely cycle time, initial investment cost, labour requirements, line balancing, final product quality, shop floor area consumed, lead time to manufacturing the equipment, maintenance requirements, knowledge transfer availability, and flexibility.

Cycle Time

The production of a switch consists of three main operations, namely assembly, testing and finishing. The production cycle time corresponds to the longest cycle time from among these three operations. In order to derive the individual cycle time of each operation, the expected duration of every manufacturing step for each of the proposed concepts was estimated, and the operation cycle time is then given by the longest duration from among its individual stations, taking into account also the indexing time where applicable. It was found that the bottleneck of the production line for all four concepts is the assembly operation. The results of the analysis are summarized in Table 11.

Table 11: Production cycle times for each switch family under each concept

	Single gang	Old three gang	New three gang
Concept 1	6.0s	15.0s	6.0s
Concept 2	8.5s	15.0s	10.8s
Concept 3	8.5s	15.0s	6.0s
Concept 4	6.0s	15.0s	6.0s

Initial Investment Costs

The initial costs associated with each concept were estimated. It was found that Concept 4 would be the most expensive to implement, followed by Concept 1, Concept 3, and Concept 2 respectively.

Labour Requirements

The total number of human operators required for each conceptual approach was determined. Based on the descriptions given in section 4.8, Concept 1 would require nine operators, Concept 2 would require twelve operators, Concept 3 would require ten operators, and Concept 4 would require nine operators.

Line Balancing

The balance efficiency of a production line is a measure of the time used for productive work at each station as compared to the total available time (e.g. Groover, 2001). In a perfectly balanced line (i.e. 100% balance efficiency) the durations of the jobs carried out at each individual station (be they manual or automated) would be exactly equal to each other (and equal to the cycle time), and there would be no idle time at any station. In practice this ideal situation is very difficult to obtain since it is unlikely that the total work required can be broken down into discrete steps of exactly the same duration, however it remains important to strive for as high an efficiency as possible. The achievable line balance efficiencies for the four concepts described in section 4.8 are given in Table 12.

Table 12: Balance efficiencies for the single gang and new three gang switch families under each concept

	Single gang	New three gang
Concept 1	63%	68%
Concept 2	60%	59%
Concept 3	60%	68%
Concept 4	68%	

Final Product Quality

The increase in production rate should not be achieved at the expense of reduced product quality and therefore all concepts considered were developed with great concern towards maintaining high quality standards. Thus for example, the proposed automatic handling of buttons would be performed without any part coming into contact with the surface of the button. Given that a well designed automation system is often capable of greater consistency than a system based on human operators, it would be expected that product quality would increase through greater use of automation.

Shop Floor Area Consumed

Due to limitations in the shop floor area, the space consumed by the manufacturing lines would need to be minimized. It was estimated that approximately 55 m^2 of floor area would be consumed by the cells under Concept 1, 50 m^2 under Concept 2, 50 m^2 under Concept 3, and 90 m^2 under Concept 4.

Lead Time to Manufacturing the Equipment

The development of a new production line involves the mechanical and electrical systems design, manufacture of the required components, wiring and programming of the system, and assembly, tuning, and testing of the system. It was estimated that for Concept 1 the total lead time would be approximately 28 weeks, for Concept 2 it would be 2.5 weeks, for Concept 3 it would be 23 weeks, and for Concept 4 it would be 29 weeks.

Maintenance Requirements

The selected system would require both preventive and corrective maintenance tasks in order to function correctly over a period of time. The time required to perform such maintenance translates into lost production time and therefore the maintenance requirements of the developed concept need to be minimized so as to maximise productivity and efficiency. The maintenance requirements increase with the number of mechanical, pneumatic and electrical components in the system. Thus Concept 1 and Concept 4, which contain indexing tables, automated stations and vibratory bowl feeders would have higher maintenance requirements than would Concept 2 which contains only mechanized jigs.

Knowledge Transfer Availability

In today's competitive market, cost reduction and shorter lead time to market are very important. Knowledge transfer aims at achieving these goals, through sharing of the knowledge learnt from previous projects, especially in terms of technologies and procedures. The four conceptual solutions are based on layout styles that are already widely applied at the company and therefore personnel are already experienced with similar equipment. This results in a reduction of the lead time to implement the concept and an improvement in operation and troubleshooting efficiency.

Flexibility

The number of distinct members of the three switch families is expected to increase in the next few years. All four concepts that have been generated allow for these expected new variations, however future customer requirements are difficult to forecast with precision. The automation concepts that have been proposed are based on fixed automation systems and thus would not allow for major variations in switch design. An increased flexibility is however supplied by the manual assembly jigs since in the associated concepts, potential future changes to the switch design can be more easily catered for by the human operators.

The Decision Matrix

Concept selection was based on a decision matrix. The ten criteria discussed in section 4.9 were ranked in order of importance, in consultation with experienced company personnel, and were subsequently given a weighting ranging from 10 for the most important criterion, to 1 for the least important. Each concept was then assigned an individual score for each criterion, based on the analysis of section 4.9. The total score for each concept was then obtained from the weighted sum of the individual scores. The complete decision matrix is shown in Table 13. Based on this result, the selected solution was based on Concept 1.

Table 13: The decision matrix

Selection criterion	Weighting	Concept 1	Concept 2	Concept 3	Concept 4
Labour requirements	10	7	1	3	7
Cycle time	9	7	1	3	7
Final product quality	8	7	1	3	7
Initial investment cost	7	3	7	5	1
Line balancing	6	5	1	3	7
Knowledge transfer availability	5	3	3	3	3
Flexibility	4	1	5	3	1
Lead time to manufacture	3	3	7	5	1
Shop floor area consumed	2	5	7	7	1
Maintenance requirements	1	3	7	5	3
TOTAL		281	159	195	265

Process Failure Modes and Effects Analysis

A process failure modes and effects analysis (PFMEA) is a detailed analysis of the errors and malfunctions that can occur during an engineering process, including assessment of the severity, probability of occurrence, and effects of the potential malfunctions, with a view to improve the process design and reliability. An extensive PFMEA was carried out on the selected concept, searching for and assessing various potential failure modes at every station of both production cells. In addition to the various specific process design provisions that were made to address each failure mode that was identified and evaluated through the PFMEA, a number of general conclusions could be drawn from the qualitative and quantitative results of the exercise. Firstly, it was noted that various mistakes can be made by the human operators at the manual stations, and that these mistakes can be minimized by providing clear and concise working instructions to the operators. In this respect all necessary training must also be given. Secondly, it was noted that malfunction of the grippers and air supply, and errors in alignment and settings, can have significant but avoidable detrimental effects on the production process. During the PFMEA a high severity rating was assigned to all of the pick and place operations, to motivate special attention to all associated production line components during commissioning. These ratings would later need to be revised so as to reflect better the final conditions of the line. Thirdly, due to the fact that numerous variants exist for the parts being assembled, the risk of product misidentification and mixing is high. Therefore inspection tests need to be performed in order to detect this failure mode, and this can be achieved through the addition of colour sensors on the linear feeder, a camera inspection on the end of line testing, and an automatic bar code scanner. Fourthly, it is noted that as a final measure, all potential failure modes can be detected by the end of line tests being performed, thus ensuring high reliability of the final product being delivered to the customer.

Safety Analysis

In order to ensure that all safety considerations are integrated within the project as early as possible in the design process, a safety analysis was performed on the selected concept. The analysis followed the five step approach recommended by Bahr (1997) – Step 1: Define the system; Step 2: Identify the hazards; Step 3: Evaluate the hazards; Step 4: Resolve the hazards; and Step 5: Carry out follow-up activity. The system was defined (Step 1) to encompass the two production lines (Cell 1 and Cell 2). The results of Step 2 through Step 4 of the safety analysis are summarized in Table 14. Step 5 can be realized through continued regular checks to ensure: effectiveness of all safety modules; correct functionality of all emergency stop buttons; presence of all protective covers and that all covers are tightly fixed; no cutting edges have been created by wear and tear of the machine; presence of all required grounding systems; and effectiveness of the extraction system and regular filter replacement.

Ergonomic Analysis

In order to improve worker interaction with the system being operated, ergonomic principles were applied to the system design, so as to accommodate human needs. This improves operator performance and well-being, resulting in an increase in overall system performance and efficiency. The analysis has been performed on the manual workstations to ensure that the most ergonomic design is chosen. The ergonomic design specifications are based on the recommendations in Kanawaty (1992) and Wojcikiewicz (2003). The height of the seated workbench is to be set at approximately 0.72 m so as to ensure that the worker's arms are below the shoulders. The leg clearance should be approximately 0.4 m at knee level and 0.6 m for the feet, without any obstructions such as drawers, between the legs. Height adjustable chairs are to be utilised for the all manned workstations, so as to ensure that the back and neck are not inclined more than 30°. A foot rest should also be available for operators if required. All silos and trays containing assembly parts should be placed within the maximum reach area of the operator, whereas the cavity should be placed within the optimum reach area.

Table 14: (first part) Safety analysis: identification, evaluation, and resolution of hazard

Category	Hazard description	Potential causal factors	Sev.	Occ.	Hazard Resolution
Mechanical	Crushing of body part	Unexpected movement of pneumatic cylinders	II	A	Safety guards with interlocks
		Unexpected movement of electric motors on testing station	II	A	
	Operator cuts a body part	Sharp Edges on equipment	III	B	Chamfers and edge deburring
	Operator catches a body part in a pinch point	Pulleys controlling conveyor movement	II	B	Protective covers for all pulleys
	Impact	Unexpected movement of pneumatic cylinders	II	B	Safety guards with interlocks
		Unexpected movement of indexing table	II	B	
	Wrap Points	Entanglement of clothing and accessories with conveyor	III	C	Protective covers for all pulleys; emergency stops
Electrical / Electronic	Energized equipment resulting in electric shock	Improper electrical connections and wiring	I	C	Include fuses, circuit breakers, and electrical grounding; use electrical safety checklist with double-checking; enclose wiring in control box; emergency stop switches
		Poor insulation	I	D	
		Insufficient grounding	I	D	
		Inadvertent activation	I	B	

Severity key: I-Catastrophic; II-Critical; III-Marginal; IV-Negligible.
Occurrence key: A-Frequent; B-Probable; C-Occasional; D-Remote; E-Improbable

Movement of the eyes should be minimized since it takes approximately three seconds for the eyes to rotate and refocus. Therefore the buttons and the work-piece should be placed within the 15° view angle, on either side of the centreline, since this angle requires no eye movement to allow for the grabbing of the parts. Part silos and the label printer should be placed within the 35° view angle. Correct lighting should be available since this helps to reduce errors and thus improve productivity. The light intensity requirement for the operations to be performed in this case study should be about 500 lux, where one lux is given by the illumination of a surface placed one meter away from a single candle. The light should be uniformly distributed so as to avoid pronounced shadows and excessive contrasts.

Table 14: (continued) Safety analysis: identification, evaluation, and resolution of hazards

Category	Hazard description	Potential causal factors	Sev.	Occ.	Hazard Resolution
Noise/ Vibration	Permanent damage to hearing	Environmental sound level exceeds 80dBA	II	B	Pneumatic cylinders equipped with silencers.
	Personnel fatigue	Excessive vibrations to operator's workstation	III	A	No vibratory or linear feeders placed in proximity to operators' workstations.
Lasers	Eye exposure	Collimated beam direct from the laser head into the operator's eyes	II	C	Laser systems enclosed by safety guards with interlocks
	Burning of operator hands	Collimated beam direct from the laser head over the operator's hand	II	C	
	Operator inhales toxic fumes	Toxic fumes arising from burning of plastics by laser marking.	III	A	Fume extraction system

Severity key: I-Catastrophic; II-Critical; III-Marginal; IV-Negligible.
Occurrence key: A-Frequent; B-Probable; C-Occasional; D-Remote; E-Improbable

New Capacity Analysis

A detailed capacity analysis was carried out on the proposed production system, based on the assumptions made in section 4.3. The results of this analysis are summarized in Table 15. The total required output can be reached easily using the proposed system, and even in the most demanding year (Year 4) there is a substantial reserve capacity.

Table 15: Number of shifts required to cater for the projected volumes using the proposed production lines

	Number of Daily Shifts Required			Total Number of Daily Shifts Required
Fiscal Year	Single Gang	Old Three Gang	New Three Gang	
Year 1	0.15	2.55	0.00	2.70
Year 2	1.02	1.97	0.16	3.15
Year 3	1.54	1.97	0.94	4.45
Year 4	1.67	1.97	1.63	5.27

Provisional Return on Investment Analysis

In order to estimate the financial benefits that would be gained by the company upon the implementation of the proposed layouts, a provisional return on investment analysis was carried out. The operational cost savings were calculated by comparing labour costs under the present and the proposed layouts. The labour seconds required to manufacture each switch was calculated by multiplying the cycle time by the number of operators required to operate the cell. These calculations indicate substantial cost savings over the four year period, with return on the initial investment achieved in less than three years.

CONCEPTUAL DRAWINGS

While not included among the more critical procedural guidelines proposed in section 3 above, the generation of three-dimensional renditions of the conceptual design of a system helps the design team visualize the overall concept and may aid in the optimization of the spatial layout

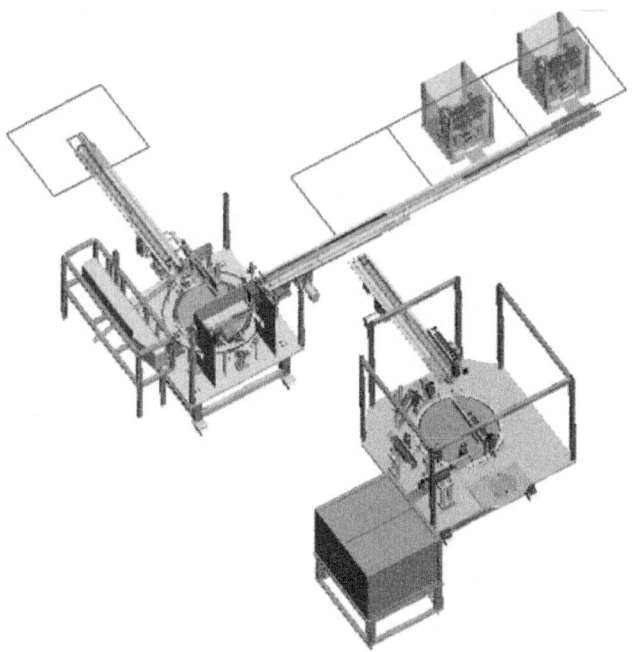

Figure. 14: A 3-D rendition of the proposed layout for Cell 1

A three-dimensional conceptual drawing of the proposed Cell 1 for this case study is given in Figure 14, and shows the two mechanized old three gang switch assembly jigs, the indexing table for single gang switch assembly, and the end of line testing module. The drawing for the proposed Cell 2 is given in Figure 15, and shows the indexing module for the assembly of the new three gang switches and the end of line testing module. These drawings were generated using Pro/ENGINEER (PTC, 2008).

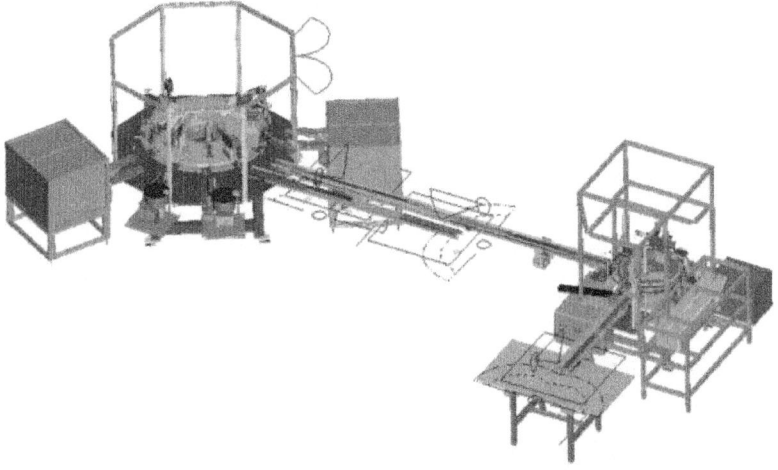

Figure. 15: A 3-D rendition of the proposed layout for Cell 2

CONCLUSION

The procedural guidelines that have been presented in this work contribute an important planning and implementation approach for the development of a conceptual design for a new manufacturing system, when migrating from manual to automated assembly of a part family of products. The novelty in the approach presented here is in the fusion of the conventional guidelines for the development of production automation systems, with a product design approach to the manufacturing system. The detailed case study that is presented in this work serves to demonstrate the application of the guidelines, and will serve as a useful reference tool for future projects of this nature. Future research in this area can include an extension of this approach to the embodiment and detailed design stages of the production system development.

In the case study it is shown that the new automated manufacturing system will result in a cycle time reduction of six seconds for the single gang switches, and of nine seconds for the new three gang switches. This will result in a corresponding increase in production capacity, thus also improving the flexibility of the company since it will be able to react to new customer orders more quickly. The new layout also results in a reduction in the manufacturing lead time, allowing the forecasted customer requirements to be catered for with over 40% excess capacity. The initial investment that is required is justified, since a significant reduction in labour costs is experienced, resulting in a return on investment of less than three years.

REFERENCES

1. Andreasen, M.M. & Hein, L. (1987). Integrated Product Development, IFS (Publications), ISBN 0948507217, Bedford, UK.

2. Asfahl, C.R. (1992). Robots and Manufacturing Automation (Second Edition), John Wiley & Sons, Inc., ISBN 0471553913, NY, USA.

3. Bahr, N.J. (1997) System Safety Engineering and Risk Assessment: A Practical Approach, CRC Press, ISBN 1560324163, London, UK.

4. Baines, T. (2004). An integrated process for forming manufacturing technology acquisition decisions. International Journal of Operations & Production Management, Vol. 24, No. 5,pp. 447–467, ISSN 0144-3577.

5. Boothroyd, G., Dewhurst, P. & Knight, W. (2001). Product Design for Manufacture and Assembly (Second Edition, Revised and Expanded), CRC Press, ISBN 082470584X,NY, USA.

6. Chan, F.T.S & Abhary, K. (1996). Design and evaluation of automated cellular manufacturing systems with simulation modelling and AHP approach: a case study. Integrated Manufacturing Systems Vol. 7, No. 6, pp. 39–52, ISSN 0957-6061.

7. Dieter, G.E. & Schmidt, L.C. (2009). Engineering Design (Fourth Edition), McGraw-Hill, ISBN 0072837039, NY, USA.

8. Groover, M.P. (2001). Automation, Production Systems, and Computer-Integrated Manufacturing (Second Edition), Prentice-Hall, ISBN 0130889784, NJ, USA.

9. Hyer, N. & Wemmerlov, U. (2002). Reorganizing the Factory: Competing through Cellular Manufacturing. Productivity Press, ISBN 1563272288, OR, USA.

10. Kanawaty, G. (ed.) (1992). Introduction to Work Study (Fourth Edition), International Labour Organization, ISBN 9221071081, Geneva, Switzerland.

11. Kapp, K.M. (1997). The USA Principle: The Key to ERP Implementation Success. APICS~The Performance Advantage, Vol. 12, pp. 62–66.

12. PTC (2008). Pro/ENGINEER design software, Parametric Technology Corporation, MA, USA. Roozenburg, N.F.M. & Eekels, J. (1995). Product Design: Fundamentals and Methods, John Wiley & Sons, Inc.,

ISBN 0471954659, NY, USA.

13. Säfsten, K., Winroth, M. & Stahre, J. (2007). The content and process of automation strategies. International Journal of Production Economics, Vol. 110, pp. 25–38, ISSN 0925-5273.

14. Ullman, D.G. (1997). The Mechanical Design Process. McGraw-Hill, ISBN 0070657564, NY, USA.

15. Winroth, M. & Säfsten, K. (2008). Automation Strategies - Implications on strategy process from refinement of manufacturing strategy content. Proceedings of the 19th Annual Conference of the Production and Operations Management Society, 008-0708, La Jolla, CA, USA, May 2008.

16. Wojcikiewicz, K. (2003). Seven Key Factors for Ergonomic Workstation Design. Manufacturing Engineering, Vol.131, No.1, pp. 45–50.

CITATION

___

CHAPTER 1

Yihai He, Zhenzhen He, Linbo Wang, and Changchao Gu, "Reliability Modeling and Optimization Strategy for Manufacturing System Based on RQR Chain," Mathematical Problems in Engineering, vol. 2015, Article ID 379098, 13 pages, 2015. doi:10.1155/2015/379098

CHAPTER 2

Jiafeng Zhang, Mohamed Khalgui, Wassim Mohamed Boussahel, et al., "Modeling and Verification of Reconfigurable and Energy-Efficient Manufacturing Systems," Discrete Dynamics in Nature and Society, vol. 2015, Article ID 813476, 14 pages, 2015 doi:10.1155/2015/813476

CHAPTER 3

R. Gopura and T. Jayawardane, "Analysis, Modeling and Simulation of a Poly-Bag Manufacturing System," Engineering, Vol. 4 No. 5, 2012, pp. 256-265. doi: 10.4236/eng.2012.45034.

CHAPTER 4

Ergün Eraslan and Berna Dengiz, "The Efficiency of Variance Reduction in

Manufacturing and Service Systems: The Comparison of the Control Variates and Stratified Sampling," Mathematical Problems in Engineering, vol. 2009, Article ID 791750, 12 pages, 2009. doi:10.1155/2009/791750.

CHAPTER 5

Xun Gong, Yixiong Feng, Hao Zheng, and Jianrong Tan, "An Adaptive Maintenance Model Oriented to Process Environment of the Manufacturing Systems," Mathematical Problems in Engineering, vol. 2014, Article ID 537452, 10 pages, 2014. doi:10.1155/2014/537452.

CHAPTER 6

Aslı Aksoy and Nursel Öztürk (2012). The Fundamentals of Global Outsourcing for Manufacturers, Manufacturing System, Dr. Faieza Abdul Aziz (Ed.), ISBN: 978-953-51-0530-5, InTech, DOI: 10.5772/36298.

CHAPTER 7

Tomasz Mączka and Tomasz Żabiński (2012). Platform for Intelligent Manufacturing Systems with Elements of Knowledge Discovery, Manufacturing System, Dr. Faieza Abdul Aziz (Ed.), ISBN: 978-953-51-0530-5, InTech, DOI: 10.5772/35095.

CHAPTER 8

Jacquelyn K. S. Nagel and Frank W. Liou (2012). Hybrid Manufacturing System Design and Development, Manufacturing System, Dr. Faieza Abdul Aziz (Ed.), ISBN: 978-953-51-0530-5, InTech, DOI: 10.5772/35597.

CHAPTER 9

Hasse Nylund and Paul H. Andersson (2012). Digital Manufacturing Supporting Autonomy and Collaboration of Manufacturing Systems, Manufacturing System, Dr. Faieza Abdul Aziz (Ed.), ISBN: 978-953-51-0530-5, InTech, DOI: 10.5772/34596.

CHAPTER 10

Michael A. Saliba and Anthony Caruana (2012). Migrating from Manual to Automated Assembly of a Product Family: Procedural Guidelines and a Case Study, Manufacturing System, Dr. Faieza Abdul Aziz (Ed.), ISBN: 978-953-51-0530-5, InTech, DOI: 10.5772/37173.

INDEX